Genetic Biotechnology

Genetic Biotechnology

P. R. Gilbert

ANMOL PUBLICATIONS PVT. LTD.
NEW DELHI - 110 002 (INDIA)

ANMOL PUBLICATIONS PVT. LTD.
H.O.: 4374/4B, Ansari Road, Darya Ganj
New Delhi-110 002 (India)
Ph.: 23278000, 23261597
B.O.: 1015, Ist Main Road, BSK IIIrd Stage
IIIrd Phase, IIIrd Block
Bangalore - 560 085 (Karnataka)
Ph: 080-41723429, Tel/Fax: 080-2672 3604
Visit us at: www.anmolpublications.com

Genetic Biotechnology

First Published, 2008

ISBN 978-81-261-3275-1

PRINTED IN INDIA

Printed at Mehra Offset Press, Delhi.

Contents

Contents

Preface

This book entitled "Genetic Biotechnology " is designed to acquaint readers with the study area of modern genetic *biotechnology* at the global level. Genetic engineering (GE) is the manipulation of genetic material (i.e. DNA or genes) in a cell or an organism in order to produce desired characteristics and to eliminate unwanted ones. GE includes a range of different techniques with many different uses, and can be applied to plants, animals and humans. For example, the genetic modification of food is a form of GE that involves manipulating the cells of plants such as maize, to increase the yields, make it more nutritious and to make it drought- and disease-resistant. However, the most contentious type of GE is definitely related to its applications in humans. GE in humans has opened up a Pandora's box of possibilities as it can be used for both the miraculous and the sinister. The cloning of Dolly the sheep in 1996 was a very important event. Until then, the cloning of a human was only possible in theory. Recent films and TV programmes such as *Gattaca, Mutant X, Dark Angel* and books such as *Brave New World* all focus on possible consequences of this technology – but what is the real deal?

Genetic engineering, genetic modification (GM), and the now-deprecated gene splicing are terms for the process of manipulating genes, usually outside the organism's normal reproductive process. It often involves the isolation, manipulation and reintroduction of DNA into cells or model organisms, usually to express a protein. The aim is to introduce new characteristics such as making a crop resistant to an herbicide, introducing a novel trait, or producing a new protein or enzyme. Examples include the production of human insulin through the use of modified bacteria, the production of erythropoietin in Chinese Hamster Ovary cells, and the production of new types of experimental mice such as the OncoMouse (cancer mouse) for research, through genetic redesign. Since a protein is specified by a segment of DNA called a gene, future versions of that protein can be modified by changing the gene's underlying DNA. One way to do this is to isolate the piece of DNA containing the gene, precisely cut the gene out, and then reintroduce (splice) the gene into a different DNA segment. Daniel Nathans and Hamilton Smith received the 1978 Nobel Prize in physiology or medicine for their isolation of restriction endonucleases, which are able to cut DNA at specific sites. Together with ligase, which can join fragments of DNA together, restriction enzymes formed the initial basis of recombinant DNA technology. Genetic engineering, the use of various methods to manipulate the DNA (genetic material) of cells to change hereditary traits or produce biological products. The techniques include the use of hybridomas (hybrids of rapidly multiplying cancer cells and of cells that make a desired antibody) to make monoclonal antibodies monoclonal antibody, an antibody that is mass produced in the laboratory from a single clone and that recognizes only one antigen. Monoclonal antibodies are typically made by fusing a normally short-lived, antibody-producing B cell to a fast-growing cell, such as a cancer cell (sometimes referred to as an "immortal" cell).; gene splicing or recombinant DNA, in which the DNA of a desired gene is inserted into the DNA of a bacterium, which then reproduces itself, yielding more of the desired gene; and polymerase chain reaction polymerase chain reaction (PCR), laboratory process in which a particular DNA segment from a mixture of DNA chains is rapidly replicated, producing a large, readily analyzed sample of a piece of DNA; the process is sometimes called DNA amplification, which makes perfect copies of DNA fragments and is used in DNA fingerprinting DNA fingerprinting or DNA profiling, any of several similar techniques for analyzing and comparing DNA from separate sources, used especially in law enforcement to identify suspects from hair, blood, semen, or other biological materials found at the scene of a violent crime.

The first clinical trial of human gene therapy began in 1990, but (as of 2006) gene therapy is still experimental. Other forms of human genetic engineering are still theoretical, or restricted to fiction stories. Recombinant DNA research is usually performed to study gene expression and various human diseases. Some drastic demonstrations of gene modification have been made with mice and other animals, however: testing on humans is generally considered off-limits. In some instances changes are usually brought about by removing genetic material from one organism, and transferring them into another species. This method is known as recombinant genetics. There are two main types of genetic engineering. Somatic modifications involve adding genes to cells other than egg or sperm cells. For example. if a person had a disease caused by a defective gene, a healthy gene could be added to the affected cells to treat the disorder. The distinguishing characteristic of somatic engineering is that it is non-inheritable, e.g. the new gene would not be passed to the recipient's offspring. Germline engineering would change genes in eggs, sperm, or very early embryos. This type of engineering is inheritable, meaning that the modified genes would appear not only in any children that resulted from the procedure, but in all succeeding generations. This application is by far the more consequential as it could open the door to the perpetual and irreversible alteration of the human species.Genetic engineering is most easily accomplished by making changes just after the egg and sperm have melded but before first division. In this way, the gene will be expressed throughout and will affect the recipients children, grandchildren, and all subsequent generations. This type of germline engineering is highly controversial and deemed inappropriate by most scientists. As of now, this is likely to take the form of gene therapy. This is less controversial as the engineered person can give consent, unlike germline engineering where subsequent generations are brought into the world with a genetic modification made to their body without their consent. Each human cell contains a different set of genes, so different genes could be transferred to different parts of the body. Such changes will not be hereditary unless the sex cells are engineered.

This book consists of—Genetically Modified Organisms: An Overview; Gene Expression, Molecular Modelling and SNPs; and Practical Aspects of Genetic Biotechnology; etc. besides a large body of list of acronyms, glossary of relevant terms, extensive bibliography, list of websites and links, for further referencing and research.

—Editor

List of Acronyms

AAI	—	American Association of Immunologists
AANP	—	American Association of Neuropathologists
AAP	—	Association of American Physicians
ABA	—	Australian Biotechnol. Assoc.
ABI	—	Applied Biosystems, Inc.
ABIC	—	Agricultural Biotechnology Intl. Conf.
ABRF	—	Association of Biomolecular Resource Facilities
ACMG	—	American College of Medical Genet.
ACS	—	American Chemical Society
ADA	—	Americans with Disabilities Act
AEC	—	Atomic Energy Commission
AES	—	American Electrophoresis Society
AFMR	—	American Federation of Medical Research
AFIP/ARP	—	Armed Forces Inst. of Pathol./Am. Registry of Pathol.
AGSG	—	Alliance of Genetic Support Groups (now Genetic Alliance)
AGT	—	Association of Genetic Technologists
AHA	—	American Heart Association
AIBS	—	Am. Inst. of Biol. Soc.
AMIA	—	American Medical Informatics Association
ANGIS	—	Australian National Genomic Info. Service
ANL	—	Argonne National Laboratory
API	—	Application Programming Interface

ARP	—	American Registry of Pathology
ASB	—	American Society for Biotechnology
ASCB	—	American Society for Cell Biology
ASCI	—	American Society for Clinical Investigation
ASHG	—	American Society for Human Genetics
ASIP	—	American Society for Investigative Pathologists
ASIS	—	American Society for Information Science
ASLME	—	American Society of Law, Medicine, and Ethics
ASM	—	American Society for Microbiology
ASPET	—	American Society for Pharmacology and Experimental Therapeutics
BER	—	Biological and Environmental Research
BIO	—	Biotechnology Industry Organization
BNL	—	Brookhaven National Laboratory
BSCS	—	Biological Sciences Curriculum Study
BS/SCF	—	Biological Sequence/Structure Computational Facility
BTCI	—	BioPharmaceutical Technology Center Institute
CATCMB	—	Center for Advanced Training in Cell and Molecular Biology
CCM	—	Chromosome Coordinating Meeting
CDC	—	Centers for Disease Control
cDNA	—	Complementary DNA
CE	—	Capillary Electrophoresis
CEPH	—	Centre d'Etude du Polymorphisme Humain
CF	—	Cystic Fibrosis
CFF	—	Cystic Fibrosis Foundation
CHH	—	Cartilage-hair Hypoplasia
CHI	—	Cambridge Healthtech Inst.

CHOP	—	Children's Hospital of Philadelphia
CIMB	—	Center for International Meeting on Biology
CIOMS	—	Council for International Organizations of Medical Sciences
CIRB	—	Colorado Institute for Research in Biotechnology
CLMA	—	Clinical Laboratory Management Association
CMT	—	Charcot-Marie-Tooth
CONTIG	—	Consortium of Teachers in Genetics
CORN	—	Council of Regional Networks for Genetic Services
CRADA	—	Cooperative Research and Development Agreement
DM	—	Myotonic Dystrophy
DMD	—	Duchenne Muscular Dystrophy
DOE	—	Department of Energy
EEOC	—	Equal Employment Opportunity Commission
EL.B.A.	—	ELectronics and Biotechnology Advanced
EMBL	—	European Mol. Biol. Lab.
EMBO	—	European Molecular Biology Organisation
EMG	—	Encyclopedia of the Mouse Genome
EMS	—	Environmental Mutagen Society
EORTC	—	European Organization for Research and Treatment of Cancer
ERDA	—	Energy Research and Development Administration
ERI	—	Eleanor Roosevelt Institute
ES	—	Embryonic Stem
ESF	—	European Science Foundation
ESHG	—	European Society of Human Genetics

ESI — Electrospray Ionization
EST — Expressed Sequence Tag
FCM — Flow Cytometry
FEBS — Fed. of Eur. Biochem. Soc.
FMF — Familial Mediterranean Fever
FRAXA — Fragile X locus
FVEA — Fundacion Valenciana de Estudios Avanzados
GBASE — Genome Database of the Mouse
GBR — Global Business Research
GDB — Genome Database
GDB/OMIM — Genome Database/Online Mendelian Inheritance in Man
GESTEC — Genome Science and Technology Center
GIST — Genome Informatics System of Transputers
GLaRGG — Great Lakes Regional Genetics Group
GMCRF — General Motors Cancer Res. Foundation
GMD — Genomic Map Design
GPI — Genetics and Public Issues Program
GRAIL — Gene Recognition and Analysis Internet Link
GRC — Gordon Res. Conf.
HERAC — Health and Environmental Research Advisory Committee
HGCC — Human Genome Coordinating Committee
HGM — Human Genome Meeting
HGMIS — Human Genome Management Information System
HGP — Human Genome Project
HHMI — Howard Hughes Medical Institute
HICSS — Hawaii Intl. Conf. on Systems Sci.
HLA — Human Leukocyte Antigen

HMDP	—	Homology Database
HNPCC	—	Hereditary Nonpolyposis Colorectal Cancer
IBC	—	International Business Communications
IBEX	—	International Biotechnology EXpo
IBF	—	International Business Forum
IBI	—	Institute for Biotechnology Information
ICES	—	International Council of Electrophoresis Societies
ICGEB	—	International Centre for Genetic Engineering and Biotechnology
ICHG	—	International Congress of Human Genetics
ICPEMC	—	International Commission on Protection Against Environmental Mutagens and Carcinogens
ICRF	—	Imperial Cancer Research Fund
IEEE	—	Institute of Electrical and Electronics Engineers
IG	—	IntelliGenetics
IGES	—	International Genetics Epidemiology Societies
IJCAI	—	International Joint Conference on Artificial Intelligence
IMA	—	Institute for Mathematics and its Applications
IMACS	—	International Association for Mathematics and Computers In Simulation
IMAGE	—	Integrated Molecular Analysis of Gene Expression
IMGS	—	Intl. Mammalian Genome Society
INRIA	—	French Natl. Inst. for Research in Computer Science and Control
IOM	—	Institute of Medicine

IOR	—	Institute of Religion
ISAG	—	Intl. Society for Animal Genetics
ISMB	—	Intelligent Systems for Molecular Biology
ISONG	—	Intl. Soc. of Nurses in Genet.
ISQL	—	Interactive Standard Query Language
ISTR	—	Institute for Science Training and Research
IUBMB	—	Intl. Union of Biochemistry and Molecular Biology
IU	—	Indiana University
LBNL	—	Lawrence Berkeley National Laboratory
LLNL	—	Lawrence Livermore National Laboratory
LTI	—	Life Technologies, Inc.
MBC	—	Massachusetts Biotechnology Council
MBL	—	Marine Biological Laboratory
MCD	—	Mouse Cytogenetic Database
MDA	—	Muscular Dystrophy Assoc.
MEMS	—	Microelectromechanical Systems
MGC	—	Mouse Genome Conference
MGD	—	Mouse Genome Database
MGI	—	Microbial Genome Initiative
MHC	—	Major Histocompatibility Complex
MIMBD	—	Meet. on the Interconnection of Mol. Biol. Databases
MIT	—	Massachusetts Institute of Technology
MMRF	—	Marshfield Medical Research Foundation
MNBWS	—	Miami Nature Biotechnology Winter Symposium
MOD	—	March of Dimes
MOU	—	Memorandum of Understanding
MRC	—	Medical Research Council
NAPBC	—	Natl. Action Plan on Breast Cancer

NAS	—	Natl. Academy of Sciences
NASA	—	Natl. Aeronautics and Space Administration
NBAC	—	Natl. Bioethics Advisory Commission
NCGR	—	National Center for Genome Resources
NCI	—	Natl. Cancer Institute
NCSA	—	National Center for Supercomputing Applications
NCSL	—	National Conference of State Legislatures
NCSU	—	North Carolina State University
NEH	—	National Endowment for the Humanities
NFCR	—	National Foundation for Cancer Research
NFID	—	National Foundation for Infectious Diseases
NHGRI	—	Natl. Human Genome Res. Inst.
NICHD	—	National Institute of Child Health and Human Development
NIGMS	—	National Institute of General Medical Sciences
NIH	—	National Institutes of Health
NIGMS	—	Natl. Institute of General Medical Sciences
NIST	—	Natl. Inst. of Standards and Technology
NLGLP	—	National Laboratory Gene Library Project
NLM	—	National Library of Medicine
NMHCC	—	National Managed Health Care Congress
NORD	—	National Organization for Rare Disorders
NRC	—	National Research Council
NRSA	—	National Research Service Award
NSF	—	Natl. Sci. Foundation
NSGC	—	National Society of Genetic Counselors
OHER	—	Office of Health and Environmental Research

OMIM	—	Online *Mendelian Inheritance in Man*
OPRR	—	Office of Protection from Res. Risks
ORAU	—	Oak Ridge Associated Universities
ORISE	—	Oak Ridge Inst. for Science and Education
ORF	—	open reading frame
ORNL	—	Oak Ridge National Laboratory
OSU	—	Oregon State University
PCR	—	Polymerase Chain Reaction
PFGE	—	Pulsed-field gel electrophresis
PG	—	Plant Genome
PIR	—	Protein Info. Resource
PNNL	—	Pacific Northwest National Laboratory
PRIM&R	—	Public Responsibility in Medicine & Research
RAPD	—	Random Amplified Polymorphic DNA
RARA	—	Retinoic Acid Receptor
RECOMB	—	Conference on Computational Molecular Biology
RFLP	—	Restriction Fragment Length Polymorphism
SCAN	—	Sequence Comparison ANalysis Program
SCE	—	School of Continuing Education
SCI	—	Society of Chemical Industry
SCW	—	Single-chromosome Workshop
SDC	—	San Diego Conference
SERGG	—	Southeast Regional Genetics Group
SHOM	—	Sequencing by Hybridization on Matrices
SIAM	—	Society for Industrial and Applied Mathematics
SIB	—	Society for Industrial Biology
SIM	—	Society for Industrial Microbiology

SIMS	—	Societal Institute of the Mathematical Sciences
SIVB	—	Society for In Vitro Biology
SNL	—	Sandia National Laboratory
SNP	—	Single Nucleotide Polymorphism
SQL	—	Standard Query Language
SSLP	—	Single-sequence Length Polymorphism
SSR	—	Simple Sequence Repeat
STM	—	Scanning Tunneling Microscope
STRP	—	Short Tandem Repeat Polymorphism
TIGR/NIST	—	The Inst. for Genomic Res./Natl. Inst. of Standards and Technol.
UCB	—	University of California, Berkeley
UCLA	—	University of California, Los Angeles
UICM	—	Univ. of Illinois College of Medicine
UMBC	—	University of Maryland, Baltimore County
UNC	—	University of North Carolina at Chapel Hill
UMDS	—	United Medical & Dental Schools (Univ. of London)
UNESCO	—	United Nations Educational, Scientific, and Cultural Organization
USDA	—	U.S. Dept. of Agriculture
UT	—	University of Tennessee
WHS	—	Wolf-Hirschhorn syndrome
WLMG	—	Wellcome Laboratory for Molecular Genetics
WSES	—	World Scientific and Engineering Society

List of Glossary of Terms

Antibiotic resistance: The ability of a microorganism to produce a protein that disables an antibiotic or prevents transport of the antibiotic into the cell.

Anticodon: A nucleotide base triplet in a transfer RNA molecule that pairs with a complementary base triplet, or codon, in a messenger RNA molecule. See Codon, Messenger RNA, RNA.

Antigenic switching: The altering of a microorganism's surface antigens through genetic rearrangement, to elude detection by the host's immune system.

Bacteriocide: A class of antibiotics that kills bacterial cells.

Bacteriophage (phage or phage particle): A virus that in- fects bacteria. Altered forms are used as vectors for cloning DNA.

Bacteriorhodopsin: Pigmented protein found in the plasma membrane of a salt-loving bacterium, *Halobacterium halobium*; it pumps protons out of the cell in response to light.

Basal body: Short cylindrical array of microtubules plus their associated proteins found at the base of a eucaryotic cell cilium or flagellum. Serves as a nucleation site for the growth of the axoneme. Closely similar in structure to a centriole.

Basal lamina (plural basal laminae): Thin mat of extracellular matrix that separates epithelial sheets, and many types of cells such as muscle cells or fat cells, from connective tissue. Sometimes called a basement membrane.

Beta-DNA: The normal form of DNA found in biological systems, which exists as a right-handed helix.

Biomotors: Driven by energy sources such as *adenosine triphosphate (ATP)* for chemical transduction and other processes. These biomotors are considered to be biomolecular and are discussed in the body of this report, but strictly speaking they do not conform to the panel's definition of self- assembly.

Bioremediation: The use of microorganisms to remedy environmental problems. See Bioaugmentation, Bioenrichment.

Biosensor: The term "biosensor" is a general designation that denotes either a sensor to detect a biological substance or a sensor which incorporates the use of biological molecules such as antibodies or enzymes. Biosensors are a subcategory of chemical sensors.

Biotechnology: The scientific manipulation of living organisms, especially at the molecular genetic level, to produce useful products. Gene splicing and use of recombinant DNA (rDNA) are major techniques used. Technology for working with biological systems. Includes genetic engineering, human and veterinary medicine, crop and animal breeding, diagnostics, pharmaceuticals, forensics, etc. Narrow sense: Genetic engineering.

Biotic stress: Living organisms which can harm plants , such as viruses, fungi, and bacteria, and harmful insects. See Abiotic stress.

Biovorous: From "biovore;" an organism capable of converting biological material into energy for sustenance.

Bipolar-junction transistor: Transistor with n-type and p-type semiconductors having base-emitter and collector-base junctions.

cDNA: DNA synthesized from an RNA template using reverse transcriptase. Complementary DNA to a particular RNA fragment.

Cell pharmacology: Delivery of drugs by medical nanomachines to exact locations in the body.

Centromere: The central portion of the chromosome to which the spindle fibers attach during mitotic and meiotic division.

Chemotherapy: A treatment for cancers that involves administering chemicals toxic to malignant cells.

Chloramphenicol: An antibiotic that interferes with protein synthesis.

Chromatid: Each of the two daughter strands of a duplicated chromosome joined at the centromere during mitosis and meiosis.

Cistron: A DNA sequence that codes for a specific polypeptide; a gene. See DNA, Gene.

Clone: An exact genetic replica of a specific gene or an entire organism. See Cloning.

Cloning vector: A DNA molecule capable of autonomous replication within the cloning host cell (e.g *E. coli*). The vectors contain restriction enzyme sites for insertion of foreign DNA. Cloning vectors are derived from bacterial plasmids, bacteriophages, or viruses.

Coat protein (capsid): The coating of a protein that enclosed the nucleic acid core of a virus.

Colony: A group of identical cells (clones) derived from a single progenitor cell.

Combinatorial materials design: Uses computing power (sometimes together with massive parallel experimentation) to screen many different materials possibilities to optimize properties for specific applications (e.g., catalysts, drugs, optical materials).

Commensalism: The close association of two or more dissimilar organisms where the association is advantageous to one and doesn't affect the other(s). See Parasitism, Symbiosis.

Competency: An ephemeral state, induced by treatment with cold cations, during which bacterial cells are capable of uptaking foreign DNA.

Concatemer: A DNA segment composed of repeated sequences linked end to end.

Conjugation: The joining of two bacteria cells when genetic material is transferred from one bacterium to another.

Coulomb [C]: Measure of electrical charge: 1 C is an amount of charge equal to that of about 6.24x1018 electrons.

Coupling: A connection between more than one object or energy pathway so that together they function as a single unit.

Crossing-over: The exchange of DNA sequences between chromatids of homologous chromosomes during meiosis.

Cross-pollination: Fertilization of a plant from a plant with a different genetic makeup.

Crosstalk: Electromagnetic noise transmitted between leads or circuits in close proximity to each other.

Crybiology: The science of biology at low temperatures; research in cryobiology has made possible the freezing and storing of sperm and blood for later use.

Crystal Lattice: The regular three-dimensional pattern of atoms in a crystal.

Crystal structure: For crystalline materials, the manner in which atoms or ions are arrayed in space. It is defined in terms of the unit cell geometry and the atom positions within the unit cell.

Crystallescence: In medical nanorobotics, the crystallization of solid solute that is offloaded by nanorobot sorting rotors at a concentration that exceeds the solvation capacity of the surrounding solvent.

Culture: A particular kind of organism growing in a laboratory medium.

Curie temperature (also Curie point) (Tc): The temperature above which a ferromagnetic or ferrimagnetic material becomes paramagnetic. For iron the Curie point is 760oC and for nickel 356oC.

Current [A]: Measure of rate of flow of electric charge: a one-ampere current is a flow of 1 C of charge per second.

Cutoff: Condition in a diode or bipolar junction transistor in which the potential across a p-n junction prevents current flow.

Cyclic AMP (cyclic adenosine monophosphate): A second messenger that regulates many intracellular reactions by transducing signals from extracellular growth factors to cellular metabolic pathways.

Diploid cell: A cell which contains two copies of each chromosome. See Haploid cell.

Double helix: Describes the coiling of the antiparallel strands of the DNA molecule, resembling a spiral staircase in which the paired bases form the steps and the sugar-phosphate backbones form the rails.

Double-stranded complementary DNA (dscDNA): A

duplex DNA molecule copied from a cDNA template.

Downstream: The region extending in a 3' direction from a gene.

Enabling science and technologies: Areas of research relevant to a particular goal, such as nanotechnology. Also, technology that "enables" other technology to advance, such as the transistor enabled the computer chip revolution, as did photolithography.

Encapsidation: Process by which a virus' nucleic acid is enclosed in a capsid. See Coat protein.

Endocrine cell: Specialized animal cell that secretes a hormone into the blood; usually part of a gland, such as the thyroid or pituitary gland.

Endocytosis: Uptake of material into a cell by an invagination of the plasma membrane and its internalization in a membrane-bounded vesicle.

Endophyte: An organism that lives inside another.

Endoplasmic reticulum (ER): Labyrinthine, membrane-bounded compartment in the cytoplasm of eucaryotic cells, where lipids are synthesized and membrane-bound proteins are made.

Energy [J]: Capacity for performing work or to cause heat flow. Like work itself, it is measured in Joules.

Environmental Protection Agency (EPA): The U.S. regulatory agency for biotechnology of microbes. The major laws under which the agency has regulatory powers are the Federal Insecticide, Fungicide, and Rodenticide Act (FIFRA); and the Toxic Substances Control Act (TSCA).

Enzymes: Proteins that control the various steps in all chemical reactions.

Ethidium bromide: A fluorescent dye used to stain DNA and RNA. The dye fluoresces when exposed to UV light.

Eukaryote: An organism whose cells possess a nucleus and other membrane-bound vesicles, including all members of the protist, fungi, plant and animal kingdoms; and excluding viruses, bacteria, and blue-green algae. See Prokaryote.

Eurisko: A computer program developed by Professor Douglas Lenat which is able to apply heuristic rules for performing various tasks, including the invention of new heuristic rules.

Exon: A DNA sequence that is ultimately translated into protein.

Express: To translate a gene's message into a molecular product.

Face-centered cubic: A crystal structure found in common elemental metals also FCC. Within the cubic unit cell, atoms are located at all corner and face-centered positions.

Fail-stop: Describes a component or subsystem that, in the event of a failure, produces no output (e.g., of material or data) rather than producing a damaged or incorrect output.

Fermi energy: The energy level in a solid at which the probability of finding an electron is 1/2. For a metal, the energy corresponding to the highest filled electron state in the valence band at 0 K.

Field-Effect Transistor (FET): Field-Effect Transistor — semiconductor device whose insulated gate electrode controls current flow.

FIFRA: The Federal Insecticide, Fungicide, and Rodenticide Act. See Environmental Protection Agency.

Fingerprint: A set of molecular markers sufficiently diverse to identify particular individuals with reasonable certainty.

Flanking region: The DNA sequences extending on either side of a specific locus or gene.

Food and Drug Administration (FDA): The U.S. agency responsible for regulation of biotechnology food products. The major laws under which the agency has regulatory powers include the Food, Drug, and Cosmetic Act; and the Public Health Service Act.

Fusion gene: A hybrid gene created by joining portions of two different genes (to produce a new protein) or by joining a gene to a different promoter (to alter or regulate gene transcription).

Gamete: A haploid sex cell, egg or sperm, that contains a single copy of each chromosome.

Ganglioside: Any glycolipid having one or more sialic acid residues in its structure. Found in the plasma membrane of eucaryotic cells and especially abundant in nerve cells.

Gel, agarose: A substance used to separate DNA or RNA fragments by size. Used for southern and northern blots.

Gel, polyacrylamide: A substance used to separate proteins by size. Used for western blots.

GEM: A genetically engineered microorganism.

Gene amplification: The presence of multiple genes. Amplification is one mechanism through which proto-oncogenes are activated in malignant cells.

Gene cloning: The process of synthesizing multiple copies of a particular DNA sequence using a bacteria cell or another organism as a host. See DNA, Host.

Gene expression: The process of producing a protein from its DNA- and mRNA-coding sequences.

Gene flow: The exchange of genes between different but (usually) related populations.

Gene frequency: The percentage of a given allele in a population of organisms. See Allele.

Gene gun: Transformation technique that uses accelerated particles coated with DNA to introduce foreign DNA into recipient plant.

Gene insertion: The addition of one or more copies of a normal gene into a defective chromosome.

Gene linkage: The hereditary association of genes located on the same chromosome.

Gene modification: The chemical repair of a gene's defective DNA sequence. See DNA.

Gene pool: The totality of all alleles of all genes of all individuals in a particular population.

Genetic drift: Random variation in gene frequency from one generation to another.

Genetic engineering: Genetic engineering is the process of manually adding new DNA on a molecular level with the goal of adding one or more new traits that are not already found in that organism. For the purposes of this web site, genetic engineering includes recombinant DNA and gene splicing technologies.

Genetic marker: A gene or group of genes used to "mark" or track the action of microbes.

Genome: The genetic complement contained in the chromosomes of a given organism, usually the haploid chromosome state. The full chromosome set containing all the genes of a particular individual.

Genomics:: The study of the structure and function of genomes. Genomics usually involves high speed sequencing of the DNA and computer searches for sequences that code for genes.

Genotype: The sum total of the genetic information of an organism including the linkage relationships between genes. The genotype, modified by environment, determines the phenotype.

Genus: A category including closely related species. Interbreeding between organisms within the same category can occur.

Germ cell (germ line) gene therapy: The repair or replacement of a defective gene within the gamete-forming tissues, which produces a heritable change in an organism's genetic constitution.

Green revolution: Advances in genetics, petrochemicals, and machinery that culminated in a dramatic increase in crop productivity during the third quarter of the 20th century.

GUI: Graphical User Interface — hardware, software, and firmware that produces the display on modern personal computers.

Guy Fawkes Scenario: If nanotechnology becomes widely available, it might become trivial for anyone to committ acts of terrorism (such as making nanomachines build a large amount of explosives under government buildings a la Guy Fawkes). This would either force strict control over nanotechnology (hard) or a decentralized mode of organization.

Hair cell: Specialized sensory epithelial cell in the ear with bundles of giant microvilli (stereocilia) protruding from its apical surface. Sound vibrations tilt the stereocilia, evoking an electrical change in the hair cell, which thus acts as a sound detector.

Hall effect: The phenomenon whereby a force is applied to a moving electron or hole by a magnetic field that is applied perpendicular to the direction of motion. The force direction is perpendicular to both the magnetic field and the particle motion directions.

Haploid cell: A cell containing only one set, or half the usual (diploid) number, of chromosomes.

Heme: An iron complex. Cyclic organic molecule containing an iron atom that carries oxygen in hemoglobin and carries an electron in cytochromes.

Heterochromatin: Dark-stained regions of chromosomes thought to be for the most part genetically inactive.

Heteroduplex: A double-stranded DNA molecule or DNA-RNA hybrid, where each strand is of a different origin.

Homologous chromosomes: Chromosomes that have the same linear arrangement of genes—a pair of matching chromosomes in a diploid organism. See Chromosomes.

Homologous recombination: The exchange of DNA fragments between two DNA molecules or chromatids of paired chromosomes (during crossing over) at the site of identical nucleotide sequences.

Homozygote: An organism whose genotype is characterized by two identical alleles of a gene. See Allele, Genotype.

Host: An organism that contains another organism.

Human growth hormone (HGH, somatotrophin): A protein produced in the pituitary gland that stimulates the liver to produce somatomedins, which stimulate growth of bone and muscle.

Hydrogen bond: A relatively weak bond formed between a hydrogen atom (which is covalently bound to a nitrogen or oxygen atom) and a nitrogen or oxygen with an unshared electron pair.

Hydrolysis: A reaction in which a molecule of water is added at the site of cleavage of a molecule to two products.

In situ: Refers to performing assays or manipulations with intact tissues.

In vivo: Refers to biological processes that take place within a living organism or cell.

Incomplete dominance: A condition where a heterozygous off- spring has a phenotype that is distinctly different from, and intermediate to, the parental phenotypes. See Heterozygote, Phenotype.

Initiation codon: The mRNA sequence AUG, coding for methionine, which initiates translation of mRNA.

insect resistance management: A set of strategies to reduce the frequency of resistance in crop pests. These strategies are widely referred to as Insect Resistance Management.

Inscrtion mutations: Changes in the base sequence of a DNA molecule resulting from the random integration of DNA from another source. See DNA, Mutation.

Insulator: A substance, object or material that does not conduct electricity. Material that conducts electricity very poorly.

Insulin: A peptide hormonc secreted by the islets of Langerhans of the pancreas that regulates the level of sugar in the blood.

Interferon: A family of small proteins that stimulate viral resistance in cells.

Intergenic regions: DNA sequences located between genes that comprise a large percentage of the human genome with no known function.

Introgression: Backcrossing of hybrids of two plant populations to introduce new genes into a wild population.

Intron: A noncoding DNA sequence within a gene that is initially transcribed into messenger RNA but is later snipped out. See Coding, DNA, Messenger RNA, Transcription.

Invasiveness: Ability of a plant to spread beyond its introduction site and become established in new locations where it may provide a deliterious effect on organisms already existing there.

Kanamycin: An antibiotic of the aminoglycoside family that poisons translation by binding to the ribosomes.

Karyotype: All of the chromosomes in a cell or an individual organism, visible through a microsope during cell division.

Kb (kilobase): The number of base pairs (in thousands) that denote the size of a nucleic acid fragment.

Kilohertz (kHz): One thousand cycles per second.

Kilojoule: Standard unit of energy equal to 1000 joules, or 0.24 kilocalories.

Lag phase: The initial growth phase, during which cell number remains relatively constant prior to rapid growth. See growth phase.

Lane: On a gel or a blot, a lane belongs to one sample. It contains fragments of different sizes that were separated by gel electrophoresis. See band, blot.

Lawn: A uniform and uninterrupted laver of bacterial growth, in which individual colonies cannot be observed.

Lectin: Protein that binds tightly to a specific sugar. Abundant lectins derived from plant seeds are often used as affinity reagents to purify glycoproteins or to detect them on the surface of cells.

Legume: A member of the pea family that possesses root nodules containing nitrogen-fixing bacteria.

Library, DNA: A large set of clones of DNA fragments from a particular organism. DNA libraries are usually maintained in *E. coli* or in bacteriophages.

Life (lifetime): The length of time the sensor can be used before its performance changes.

Ligase (DNA ligase): An enzyme that catalyzes a condensation reaction that links two DNA molecules via the formation of a phosphodiester bond between the 3' hydroxyl and 5' phosphate of adjacent nucleotides.

Ligate: The process of joining two or more DNA fragments.

Ligation: Joining two fragments of DNA end to end. See sticky ends.

Linked genes/markers: Genes and/or markers that are so closely associated on the chromosome that they are coinherited in 80% or more of cases.

Linker: A short, double-stranded oligonucleotide containing a restriction endonuclease recognition site, which is ligated to the ends of a DNA fragment.

Locus (plural = loci): A specific location or site on a chromosome.

Logarithmic phase (log or exponential growth phase): The steepest slope of the growth curve—the phase of vigorous growth during which cell number doubles every 20-30 minutes.

Lysogen: A bacterial cell whose chromosome contains integrated viral DNA.

Lytic: A phase of the virus life cycle during which the virus replicates within the host cell, releasing a new generation of viruses when the infected cell lyses.

Machine-phase chemistry: The chemistry of systems in which all potentially reactive moieties follow controlled trajectories (e.g., guided by molecular machines working in vacuum).

Macroscale: Larger than nanoscale; often implies a design that humans can directly interact with; too large to be built by a single assembler (one cubic micron of diamond contains 176 billion atoms).

Magnetostrictive material: A material that changes dimension in the presence of a magnetic field or generates a magnetic field when mechanically deformed.

Malignant: Having the properties of cancerous growth.

Manufacturing: The ability to make products, in this case ranging from clothing, to electronics, to medical devices, to books, to building materials, and much more.

Meat Machine: AKA Cabinet Beast. A box containing assemblers and raw material, within which is formed meat [or whatever else it was programmed to make].

Megabyte (MB): 220 (= 1,048,576, or about one million) bytes of information.

Megahertz (Mhz): One million cycles per second

Meiosis: The reduction division process by which haploid gametes and spores are formed, consisting of a single duplication of the genetic material followed by two mitotic divisions.

Metabolism: The biochemical processes that sustain a living cell or organism.

Metallothionein: A protective protein that binds heavy metals, such as cadmium and lead.

Micelles: Micelles are small, spherical structures composed of molecules that attract one another to reduce surface tension. The head of the molecule is hydrophilic, meaning it likes water, while the interior portion is hydrophobic, meaning it avoids water.

Microbial mats (biofilms): Layered groups or communities of microbial populations.

Microelectronics: Narrower terms: MEMS, nanoelectronics, optoelectronics, SED Single Electron Devices. Related terms: molecular electronics, semiconductors

Microencapsulation: Individually encapsulated small particles.

Microinjection: A means to introduce a solution of DNA, protein, or other soluble material into a cell using a fine microcapillary pipet.

Microsystem: A microscale machine that can sense information from the environment and act accordingly. Outside the U.S., it can also refer to microelectromechanical systems (MEMS). [smalltimes glossary, 2002]

Microtubule: A hollow cylindrical tube used for the transport of materials inside cells.

Miller indices: A set of 3 integers (4 for hexagonal) that designate crystallographic planes, as determined from reciprocals of fractional axial intercepts.

Millimeter: One thousandth of a meter, or about 1/26 of an inch.

Mitosis: The replication of a cell to form two daughter cells with identical sets of chromosomes.

Molecular - Nanotechnology: The technology of precisely-constructed molecular-scale machines; from nanometer: a billionth of a meter. For a good introduction, see What is Nanotechnology? or Ralph C. Merkle's nanotechnology page.

Molecular Beam Epitaxy: [MBE] Process used to make compound (multi-layer) semiconductors. Consists of depositing alternating layers of materials, layer by layer, one type after another (such as the semiconductors gallium arsenide and aluminum gallium arsenide).

Molecular biology: The study of the biochemical and molecular interactions within living cells.

Molecular genetics: The study of the flow and regulation of genetic information between DNA, RNA, and protein molecules.

Molecular markers: Genetic traits which are detectable on gels or blots and can be used to construct genetic maps. Molecular markers usually have no known function. RFLPs are molecular markers.

Molecular Wire: A molecular wire—the simplest electronic component—is a quasi-one-dimensional molecule that can transport charge carriers (electrons or holes) between its ends. [Michael D Ward]

Monoclonal antibodies: Immunoglobulin molecules of single- epitope specificity that are secreted by a clone of B cells.

Monoculture: The agricultural practice of cultivating crops consisting of genetically similar organisms.

Monogenic: Controlled by or associated with a single gene.

Mycorrhizae: Fungi that form symbiotic relationships with roots of more developed plants.

Nanoscience: Nanoscience is the science of nanoscale phenomena. This is a very complicated area, since Quantum Mechanics is often needed to describe the physical properties of atoms and molecules properly.

Nitrocellulose: A membrane used to immobilize DNA, RNA, or protein, which can then be probed with a labeled sequence or antibody.

Nitrogen fixation: The conversion of atmospheric nitrogen to biologically usable nitrates.

Occupational Safety and Health Administration (OSHA): One of the U.S. agencies responsible for regulation of biotechnology. The major law under which the agency has regulatory powers is the Occupational Safety and Health Act.

Oncogenesis: The progression of cytological, genetic, and cellular changes that culminate in a malignant tumor.

Open pollination: Pollination by wind, insects, or other natural mechanisms.

Operator: A prokaryotic regulatory element that interacts with a repressor to control the transcription of adjacent structural genes.

Organelle: A cell structure that carries out a specialized function in the life of a cell.

Origin of replication: The nucleotide sequence at which DNA synthesis is initiated.

Paleontology: The study of the fossil record of past geological periods and of the phylogenetic relationships between ancient and contemporary plant and animal species.

Parasitism: The closee association of two or more dissimilar organisms where the association is harmful to at least one. See Commensalism, Parasitism, Symbiosis.

Particle bombardment: Using metal particles to blast DNA into cells for the purpose of genetic engineering. A gene gun is used to propel the metal particles.

Pheromone: A hormone-like substance that is secreted into the environment.

Phosphatase: An enzyme that hydrolyzes esters of phosphoric acid, removing a phosphate group.

Phosphodiester bond: A bond in which a phosphate group joins adjacent carbons through ester linkages. A condensation reaction between adjacent nucleotides results in a phosphodiester bond between 3' and 5' carbons in DNA and RNA.

Phosphorylation: The addition of a phosphate group to a compound.

Piezoelectric material: A ferroelectric material in which an electrical potential difference is created due to mechanical deformation, or conversely, in which the application of a voltage causes dimensional changes in the material.

Population: A local group of organisms belonging to the same species and capable of interbreeding.

Primary cell: A cell or cell line taken directly from a living organism, which is not immortalized.

Primer: A short DNA or RNA fragment annealed to single-stranded DNA, from which DNA polymerase extends a new DNA strand to produce a duplex molecule.

Prokaryote: A bacterial cell lacking a true nucleus; its DNA is usually in one long strand. See Eukaryote.

Pronucleus: Either of the two haploid gamete nuclei just prior to their fusion in the fertilized ovum.

Protease: An enzyme that cleaves peptide bonds that link amino acids in protein molecules.

Protein kinase: An enzyme that adds phosphate groups to a protein molecule at serine, threonine, or tyrosine residues.

Recessive gene: Characterized as having a phenotype expressed only when both copies of the gene are mutated or missing.

Recessive: Refers to the member of a pair of alleles that fails to be expressed in the phenotype of the organism when the dominant member is present. Also refers to the phenotype of an individual that has only the recessive allele.

Rhizobia: Bacteria in a symbiotic relationship with leguminous plants that results in nitrogen fixation. See Nitrogen fixation.

Ribosome: Cellular organelle that is the site of protein synthesis during translation. See Organelle, Translation.

Ribosome-binding site: The region of an mRNA molecule that binds the ribosome to initiate translation.

Risk assessment: Defined as a formalized basis for the objective evaluation of risk in a manner in which assumptions and uncertainties are clearly considered and presented.

RNA: Ribonucleic acid. The three basic types are messenger RNA (mRNA), ribosomal RNA (rRNA), and transfer RNA (tRNA).

Roundup Ready: A trademark for plants that are genetically engineered to be resistant to the herbicide Roundup (technical name: glyphosate).

Salmonella: A genus of rod-shaped, gram-negative bacteria that are a common cause of food poisoning.

Satellite RNA (viroids): A small, self-splicing RNA molecule that accompanies several plant viruses, including tobacco ringspot virus.

Self-pollination: Pollen of one plant is transferred to the female part of the same plant or another plant with the same genetic makeup.

Semiconservative replication: During DNA duplication, each strand of a parent DNA molecule is a template for the synthesis of its new complementary strand. Thus, one half of a preexisting DNA molecule is conserved during each round of replication.

Sequence hypothesis: Francis Crick's seminal concept that genetic information exists as a linear DNA code; DNA and protein sequence are colinear.

Sequence: Noun: the particular order of nucleotides in a DNA or RNA fragment. Verb: to determine the particular order of nucleotides in a strand of DNA.

Sequence-tagged site (STS): A unique (single-copy) DNA sequence used as a mapping landmark on a chromosome.

Sexual reproduction: The process where two cells (gametes) fuse to form one hybrid, fertilized cell. See Asexual reproduction, Gamete, Hybrid.

Signal transduction: The biochemical events that conduct the signal of a hormone or growth factor from the cell exterior, through the cell membrane, and into the cytoplasm. This involves a number of molecules, including receptors, pro- teins, and messengers.

Site-directed mutagenesis: The process of introducing spe- cific base-pair mutations into a gene.

Species: A classification of related organisms that can freely interbreed.

Spore: A form taken by certain microbes that enables them to exist in a dormant stage. It is an asexual reproductive cell. See Asexual reproduction, Dormant.

Structure-functionalism: The scientific tradition that stresses the relationship between a physical structure and its function, for example, the related disciplines of anatomy and physiology.

Supergene: A group of neighboring genes on a chromosome that tend to be inherited together and sometimes are functionally related.

Supernatant: The soluble liquid &action of a sample after centrifugation or precipitation of insoluble solids.

Susceptibility: When individual or species will become ill or possibly die in the face of the stressful event or circumstances.

Synapsis: The pairing of homologous chromosome pairs during prophase of the first meiotic division, when crossing over occurs.

TATA box: An adenine- and thymine-rich promoter sequence located 25-30 bp upstream of a gene, which is the binding site of RNA polymerase.

T-DNA (transfer DNA, tumor-DNA): The transforming region of DNA in the Ti plasmid of Agrobacterium tumefaciens.

Template: An RNA or single-stranded DNA molecule upon which a complementary nucleotide strand is synthesized.

Terminator region: A DNA sequence that signals the end of transcription.

Tesla [T]: Unit of magnetic induction: 1T = 1 weber/m2 (also, 1T = 104 gauss).

Transgenic plant: Genetically engineered plant or offspring of genetically engineered plants. The transgenic plant usually contains material from at least one unrelated organisms, such as from a virus, animal, or other plant. See Transgenic.

Upstream: The region extending in a 5' direction from a gene.

Vaccinia: The cowpox virus used to vaccinate against smallpox and, experimentally, as a carrier of genes for antigenic determinants cloned from other disease organisms.

Variable surface glycoprotein (VSG): One of a battery of antigenic determinants expressed by a microorganism to elude immune detection.

Variation: Differences in the frequency of genes and traits among individual organisms within a population.

Vasculocyte: In medical nanorobotics, a theorized (nanorobotic) device capable of performing repairs of an injured vascular luminal surface.

Vasculography: A physical description (and mapping) of the human vascular system.

VC (Venture capitalist): A person or organisation that invests in promising companies during early development stages.

Vector: In cell biology, an agent (virus or plasmid) used to transmit genetic material to a cell or organism.

Vegetal pole: The end at which most of the yolk is located in an animal egg. The end opposite the animal pole.

Virtual Nanomedicine: Using VR to perform surgery and other functions inside the body.

Viscosity: The ratio of the magnitude of an applied shear stress to the velocity gradient that it produces; in other words: a measure of a noncrystalline material's resistance to permanent deformation.

Weediness: Unwanted effects of a plant.

Whisker: A very thin, single crystal of high perfection which has an extremely large length-to-diameter ratio. Whiskers are used as the reinforcing phase in some composites.

1

Genetically Modified Organisms: An Overview

GENETICALLY MODIFIED ORGANISM

A genetically modified organism (GMO) is an organism whose genetic material has been altered using techniques in genetics generally known as recombinant DNA technology. Recombinant DNA technology is the ability to combine DNA molecules from different sources into the one molecule in a test tube. Thus, the abilities or the phenotype of the organism, or the proteins it produces, can be altered through the modification of its genes. The term generally does not cover organisms whose genetic makeup has been altered by conventional cross breeding or by "mutagenesis" breeding, as these methods predate the discovery of the recombinant DNA techniques. Technically speaking, however, such techniques are, by definition, genetic modification. Examples of GMOs are diverse, and include transgenic experimental animals such as mice, several fish species, transgenic plants, or various microscopic organisms altered for the purposes of genetic research or for the production of pharmaceuticals. The term "genetically modified organism" does not necessarily imply, but does include, transgenic substitution of genes from another species, and research is actively being conducted in this field. For example, genes for fluorescent proteins can be co-expressed with complex proteins in cultured cells to facilitate study by biologists, and modified organisms are used in researching the mechanisms of cancer and other diseases.

The first GMO was created in 1973 by Stanley N. Cohen and Herbert Boyer, demonstrating the creation of a functional organism that combined and replicated genetic information from different species. In mid-1974, very soon after the first GMO was created, scientists called for and observed a voluntary moratorium on certain recombinant DNA experiments. One goal of the moratorium was to provide time for a conference that would evaluate the state of the new technology and the risks, if any, associated with it. That conference concluded that recombinant DNA research should proceed but under strict guidelines. Such guidelines were subsequently promulgated by the National Institutes of Health in the United States and by comparable bodies in other countries. These guidelines form the basis upon which GMOs are regulated to this day. The first transgenic animals were mice created by Rudolf Jaenisch in 1974. Jaenish successfully managed to insert foreign DNA into the early-stage mouse embryos; the resulting mice carried the modified gene in all their tissues. Subsequent experiments, injecting leukemia genes to early mouse embryos using a retrovirus vector, proved the genes integrated not only to the mice themselves, but also to their progeny.

Genetic modification involves genetic engineering, also known as *gene splicing*, a technique to splice together DNA fragments from more than one organism and thus preparing a "recombinant" DNA molecule in a test tube, producing a single piece of genetic material containing the original information from multiple fragments which can then be inserted into another organism. This is achieved by cutting up DNA molecules with restriction enzymes and splicing these fragments together using DNA ligase. A *transgenic* organism that contains such DNA sequences from a foreign organism integrated into its own genome, the term "transgenic" literally means *across gene*. A mouse or fish engineered to express the green fluorescence protein, for example, would be considered a transgenic organism, since the gene coding for the protein originated from a species of jellyfish. With current technology, transgenic organisms can be produced with only a very small proportion of extraneous DNA. For example, the genome of most mammals contains three billion basepairs of DNA, while it becomes relatively difficult to insert more than 10,000 to 20,000 basepairs of foreign DNA. More sophisticated techniques using yeast artificial chromosomes and bacterial artificial chromosomes allow insertions of up to 320,000 basepairs - approximately 0.01% of the total genome. In concept, multiple rounds of transgenesis or interbreeding of transgenics could lead to organisms with a higher proportion of foreign DNA, but cost and time considerations prevent this. In order to introduce new DNA into the receiving host, *vectors* are used. Vectors range from small circular pieces of DNA such as plasmids, to various viruses that can carry and transmit genetic information. Three processes are known by which the genetic composition of bacteria can be altered. Transformation is a process by which some bacteria are naturally capable of taking up DNA to acquire new genetic traits. This phenomenon was discovered by Frederick Griffith in 1928, although the fact that it was specifically DNA molecules that carried the genetic information was not proven until 1944. Bacteria that are competent to undergo transformation are frequently used in molecular biology. The foreign DNA uptake is facilitated by the presence of certain cations, such as Ca^{2+}, or by the use of electric current (electroporation). Transformation does not normally integrate new DNA into the bacterial chromosome. Instead, it remains on a plasmid. In conjugation, DNA is transferred from one bacterium to another via a temporary connecting tube of protein called a *pilus* (a process analogous to but biologically distinct from mating). A *plasmid* is transferred through the pilus. Conjugation is not widely used for the artificial genetic modification of bacteria, but happens often in nature. Transduction refers to the introduction of new DNA into a bacterial cell by a bacteriophage, a virus that infects bacteria. In order to gain knowledge about a particular gene's function, researchers often use ***knock out*** organisms. These organisms have a specific gene that has been functionally destroyed or "knocked out." They are used extensively in disease research with model organisms. For example, when investigating the cause of cystic fibrosis, researchers identified the CFTR gene as a likely candidate for the disease, found the mouse equivalent, bred a mouse with this gene "knocked out", and noted that the knockout mouse also had cystic fibrosis.

Like bacteria and plants, animals can be genetically modified by viral infection. However, the genetic modification occurs only in those cells that become infected, and in most cases these cells are eventually eliminated by the immune system. In some cases it is possible to use the gene-transferring ability of viruses for gene therapy, i.e. to correct diseases caused by a defective gene by supplying a normal copy of the gene. Permanent genetic modification of entire animals can be accomplished in mice. The process begins by first genetically modifying a mouse embryonic stem cell. This is normally done by physically introducing into the cell a plasmid that can integrate into the genome by a process known as transfection. During transfection the DNA integrates into the animal genome via non-homologous recombination. This altered cell is implanted into a blastocyst (an early embryo), which is then implanted into the uterus of a female mouse. A pup born from this blastocyst will be a chimera containing some cells derived from the unmodified cells of the blastocyst and some derived from the modified stem cell. By selecting mice whose germ cells (sperm- or egg-producing cells) developed from the modified cell and interbreeding them, pups that contain the genetic modification in all of their cells will be born. Baylor College of Medicine currently has one of the largest transgenic mice facilities in the country. There has also been the genetically manipulated bull Herman with 55 offspring. A human gene was built into his genetic code while in an early embryonic stage in 1990. As a result, milk from his female descendants contained the human protein lactoferrine, that can be used as medicine, but it was present at such low levels that it was not profitable to extract them. Insects can be genetically modified by injecting them with artificial transposons and a source of the enzyme transposase. The transposon, which can include new genes, is then integrated into the genome. Such insertions are unstable and can 'jump-out' in the presence of transposase. Transgenic fish are often created by microinjection. First generation is mosaic but several lines have been produced with the transgene incorporated into the germ line and transgenic fish can then be produced "the natural way" by crossing male and female gametes. Although many types of transgenic fish exists (e.g. for increased cold tolerance, antibiotic production, ornamental Glofish etc) the main focus had been on so called growth hormone transgenic fish, mainly salmonids, tilapias and carps. These fish have an over-production of growth hormone which results in increased growth rate from a few percent up to 30-40 times that of wild-types. In some species, final size is increased as well as growth rate providing an incentive for commercial breeders to farm such fish. However, ecological concerns over potential negative effects of transgenic fish in nature largely prevent the commencement of commercial production. A large and important portion of the research on transgenic fish today is therefore focusing on environmental risk-assessment of GH-transgenic fish.

Genetic modification (GM) is the subject of controversy in its own right. Some see the science itself as intolerable meddling with "natural" order, despite many known examples of natural genetic crossings occurring throughout history (see for example horizontal gene transfer). While some would like to see it banned, others push simply for required labeling of genetically modified food. Other controversies include the definition of patent and property pertaining to products of genetic engineering and the possibility of unforeseen global side effects as a result of modified organisms proliferating. The basic ethical issues involved in genetic research are discussed in the article on genetic engineering. In 2004, Mendocino County, California became the first county in the United States to ban the production of GMOs. The measure passed with a 57% majority. In 2005, a standing committee of the government of Prince Edward Island in Canada began work to assess a proposal to ban the production of GMOs in the province. This is a largely symbolic and empty gesture as PEI has already banned GMO potatoes, which account for most of its crop. In California, the Trinity and Marin counties have also imposed bans on GM crops, while ordinances to do so were unsuccessful in Butte, San Luis Obispo, Humboldt and Sonoma counties. Supervisors in the ag-rich counties of Fresno, Kern, Kings, Solano, Sutter and Tulare have passed resolutions supporting the practice. Currently, there is little international consensus regarding the acceptability and effective role of modified "complete" organisms such as plants or animals. A great deal of the modern research that is illuminating complex biochemical processes and disease mechanisms makes vast use of genetic engineering.

The practice of genetic modification as a scientific

technique is not restricted in the United States. Individual genetically modified crops (such as soybeans) are subject to intense study before being brought to market and are common in the United States, but estimates of their market saturation vary widely. Some countries in Europe have taken the opposite position, stating that genetic modification has not been proven safe, and therefore that they will not accept genetically modified food from the United States or any other country. This issue has been brought before the World Trade Organization, which determined that not allowing modified food into the country creates an unnecessary obstacle to international trade. Consequently, genetic modification within agriculture is an issue of some strong debate in the United States, the European Union, and some other countries. Some critics have raised the concern that conventionally bred crop plants can be cross-pollinated (bred) from the pollen of modified plants. Pollen can be dispersed over large areas by wind, animals, and insects. Recent research with creeping bentgrass has lent support to the concern when modified genes were found in normal grass up to 21 km (13 miles) away from the source, and also within close relatives of the same genus Agrostis. GM proponents point out that outcrossing, as this process is known as, is not new. The same thing happens with any new open-pollinated crop variety—newly introduced traits can potentially cross out into neighbouring crop plants of the same species and, in some cases, to closely related wild relatives. Defenders of GM technology point out that each GM crop is assessed on a case by case basis to determine if there is any risk associated with the outcrossing of the GM trait into wild plant populations. The fact that a GM plant may outcross with a related wild relative is not, in itself, a risk unless such an occurrence has consequences. If, for example, a herbicide resistance trait was to cross into a wild relative of a crop plant it can be predicted that this would not have any concequences except in areas where herbicides are sprayed, such as a farm. In such a setting the farmer can manage this risk by rotating herbicides. If patented genes are outcrossed, even accidentally, to other commercial fields and a person deliberately selects the outcrossed plants for subsequent planting then the patent holder has the right to control the use of those crops. This was supported in Canadian law in the case of Monsanto Canada Inc. v. Schmeiser. An often cited controversy is a hypothetical Technology Protection technology (dubbed terminator by NGOs). This yet to be commercialised technology would allow the production of first generation crops that would not generate seeds in the second generation because the plants yield sterile seeds.

The patent for this so-called "terminator" gene technology is owned by Delta and Pine Land and the USDA despite it often being mis-associated with Monsanto. In addition to the commercial protection of proprietary technology in selfpollinating crops such as soybean (a generally contentious issue) another purpose of the terminator gene is to prevent the escape of genetically modified traits from crosspollinating crops into wild-type species by sterilizing any resultant hybrids.

The terminator gene technology created a backlash amongst those who felt the technology would prevent re-use of seed by farmers growing such terminator varieties in the developing world and was ostensibly a means to exercise patent claims. Use of the terminator technology would also prevent "volunteers", or crops that grow from unharvested seed, a major concern that arose during the Starlink debacle.

Genetically modified characters, whether as heroes, villains, or backdrop, feature prominently in many works of fiction, in particular science fiction and cyberpunk, where it is used as a plot device to explain differences in a character or setting, such as explaining increased longevity or eradication of disease in a fictional civilization. In the Spiderman movie, Peter Parker was bitten by a super-spider, enhanced with the genes of many different spiders. The abilities of all these spiders were then transferred from the super-spider, into Peter, turning him into Spiderman. The videogame character Shadow the Hedgehog was originally a science experiment who was fused with the DNA of Black Doom, causing him to have the genes of aliens as well as hedgehogs. This, however, was not revealed until the game *Shadow the Hedgehog*.

THE CASE OF DUTCH GENETICALLY MODIFIED POTATOES

This report analyzes the development and introduction of genetically modified potatoes in the Netherlands. This study is part of a larger research program on "Ecological, Societal and Ethical Aspects of Biotechnology," funded by the Ministry of Agriculture, Nature Management and Fisheries. Genetically modified potatoes were chosen as a first case study for the following reasons: the potato is an important crop in Dutch agriculture; potato farmers are obliged to reduce the use of chemical pesticides, and are thus looking for alternative ways to prevent damage and loss by pests and diseases; applied research on genetically modified potatoes is relatively advanced. This report is part of a Technology Assessment (TA) process. A TA studies the impact of new technologies, and introduces the outcome of the studies in decision making regarding the new technologies. A major part of the TA is analyzing the technological development itself. The following questions are answered in this report:

- What is the state of the art in applied research on genetically modified potatoes (transgenic potatoes)?

- When shall the first transgenic potatoes be marketed in the Netherlands?
- Who are the main (Dutch) actors involved in developing and introducing transgenic potatoes? Why are they involved?
- How is the development and introduction of genetically modified potatoes related to other (structural) developments within the potato industry?
- What can be said about the economic impact of transgenic potatoes?
- What are the prospects for public acceptance of transgenic potatoes?

The information given in this report is based on documentary analysis, on data from LEI-DLO data banks, and on interviews with key persons directly or indirectly involved in the production, trade, or consumption of potatoes. The concept of the product chain is used, both for defining the actors involved, and for looking at the economic impact. The potato chain consists of all actors involved with potato research, potato breeding, potato farming, potato trade, potato processing, and potato consumption.

In 1990, almost 22,000 farms were engaged in potato production. The potato is the most important crop in arable farming, generating 45% of the total production value of all arable crops. In recent years, potato prices have decreased, mainly because of abundant supply. This has had a negative effect on income for Dutch arable farming. One of the major challenges for potato farmers is the reduction of pesticide use. The use of chemical pesticides is very high in Dutch potato cultivation, owing to intensive farming methods. In 1990, farmers used more than 12 million kg (active ingredient) of chemical pesticides for the potato crop (purchasing costs: 162 million guilders). Government environmental policies oblige farmers to reduce the use of pesticides by at least 50% by the year 2000. In 1990, Dutch potato production totaled seven million tons. This total consisted of 3.5 million tons of ware potatoes, 1 million tons of seed potatoes, and 2.5 million tons of starch potatoes. Seventy percent of all seed potatoes are exported. Of the ware potatoes, two thirds are exported - either as fresh potatoes or as potato products (frozen French fries, potato chips, etc.). In recent years, the export of potato products has shown the highest growth rate. The industrial processing of potatoes has become more and more important. More than fifty percent of all ware potatoes are now supplied to the processing industry. The trend in consumer preferences towards higher quality products and increased attention to environmental concerns is also affecting potato markets. This can be seen from the growing market for ecologically produced potatoes. The demand for higher quality products has led to "Integrated Chain Control" initiatives. This means that all stages of the potato chain are incorporated in efforts to enhance the quality of the final product. The assumption here is that each stage of production that affects the quality of the final product is determined by the preceding stage of the product chain.

Three groups of actors are directly engaged in the introduction of genetic engineering in potato breeding: public research institutes, biotechnology firms, and potato breeding companies. The most important public research institutes are the Center for Plant Breeding and Reproduction Research (CPRO-DLO) of the Ministry of Agriculture, Nature Management and Fisheries, and the Department of Plant Breeding (IVP) of the Agricultural University of Wageningen. Two new biotechnology firms are doing research on transgenic potatoes in the Netherlands: Keygene, of Wageningen, and MOGEN, of Leiden. Also important for the Dutch potato industry is the new Belgian biotechnology firm, Plant Genetic Systems (PGS). Several Dutch potato breeding companies have research contracts with PGS. The potato breeding companies, the third group of actors, are currently going through a process of restructuring. First of all, there is a concentration of companies under way, through mergers and takeovers. Second, potato breeders have expanded breeding research activities, as potato breeding research at public institutes has been restructured. Third, the rise of genetic engineering has encouraged companies to increase investments in biotechnology research, either in-house or as contract research.

All companies feel the need to stay in touch with the latest research findings. Fourth, pressure has increased on breeding companies to come up with varieties that have better pest and disease resistance, and thus need less pesticides. As a result of the restructuring process, the potato breeding companies will become stronger and more influential in the potato product chain as a whole.

Two kinds of transgenic potatoes can be distinguished: those with better resistance to pests and diseases, and those with improved characteristics for storage and processing. All important potato pests and diseases are subject to genetic engineering research, including diseases and loss caused by nematodes, viruses, fungi, bacteria, insects, and herbicides. Although several of these pests and diseases are not a major threat to potato cultivation in the Netherlands, they are important because of the large export of seed potatoes.

For instance, Dutch seed potatoes exported to Mediterranean countries are faced with subtropical diseases endemic to these countries. In the Netherlands, the first genetically modified potatoes were virus resistant potatoes.

Transgenic potatoes with resistance to Potato Virus X (PVX) will be ready for commercial introduction within a few years. Currently these potatoes are subject to cultivation tests. Genetic engineering research on resistance to nematodes and fungi - the two most important threats to the Dutch potato crop - is only in its infancy. Commercial introduction of transgenic potatoes with improved resistance to nematodes or fungi is not expected until the end of the century.

Transgenic potatoes with improved storage and processing characteristics have already been developed in the Netherlands, and are now subject to cultivation tests. In one kind of transgenic potato, the starch content has been changed. This amylose-free potato may make industrial starch processing more efficient. Another kind of transgenic potato has been made less vulnerable to bruising in order to reduce the loss of raw material during storage, transport, and processing. In Denmark, research on transgenic potatoes is directed at improving cold resistance. Potatoes with better cold resistance can be stored at lower temperatures, and therefore need less chemicals for restraining sprout growth. In other countries, genetic engineering research is directed at enhancing the starch content.

Because both the efficacy and the price of transgenic potatoes are still unknown, it is difficult to make a quantitative assessment of their economic impact. However, it is already evident that genetic engineering research requires potato breeders to increase their investments in R&D. Whether and when these investments will yield profits is still very uncertain. The necessity of making this kind of investment furthers the process of concentration among breeding companies. Return on investment in genetic engineering research is also dependent on the system of intellectual property protection.. National and international systems that pertain to plant breeding rights are currently under discussion. For potato farmers, the introduction of transgenic potatoes with improved disease resistance may lead to a shift in variable costs: lower costs for pesticides but higher costs for starting material. Transgenic potatoes with improved storage or processing characteristics may earn higher prices for potato farmers, while here too, the starting material may be more expensive. However, the economic impact can only be stated in hypothetical terms. The introduction of transgenic potatoes will reinforce the vertical integration in the potato chain. Transgenic potato varieties will only be grown under approval from the customer (trading firm, processing company, retailer or consumer). This approval will only be given if the transgenic variety has some evident qualitative improvement when compared to non-modified varieties. Qualitative improvements can involve the characteristics of both the product and the production process (for instance the environmental impact). The need for approval strengthens the integration between different stages of the product chain.

Whether transgenic potato varieties will be grown depends mostly on the attitude of the public (or consumer). As far as consumer acceptance goes, many issues are involved. The environmental impact of transgenic potatoes and their safety for human consumption are the most important. Because there are no definite findings on these issues, public attitudes remain uncertain. Public attitudes towards biotechnology have been studied in various countries, as well as in the European Community as a whole. Most studies showed that transgenic crops are valued moderately. Of course, the sine qua non is safety for human consumption and the absence of environmental effects. There seems to be a clear difference in risk perception for transgenic food products in various EC countries. Because Dutch potatoes are exported to almost all other EC member states, public attitudes in all of these countries is important for the Dutch potato industry. In an EC study on public attitudes towards biotechnology, it was shown that most people consider environmental and consumer organizations the most trustworthy sources of information on the impact of biotechnology. Concerning transgenic potatoes in the Netherlands, the environmental organizations are the most critical. They favor a restrained approach, because it is still uncertain what the long term environmental impact will be. Consumer organizations stress the importance of sufficient information for the consumer. Therefore they favor compulsory labeling. To achieve broad acceptance of transgenic agricultural products, a dialogue must be set up between proponents and critics of genetically modified food products. Only in this way may consensus be reached on which applications of genetic engineering hold benefits for all groups in society. Concerning transgenic potatoes, this implies that biotechnology firms and the potato business on the one hand, and environmental and consumer organizations on the other hand, must enter a dialogue to discuss which transgenic potatoes should be developed and introduced.

So far the development of genetically modified potatoes has been determined mainly by technological feasibility. Their commercial introduction, however, will also be determined by other structural developments in the potato industry. One of the major structural developments is horizontal and vertical integration in the potato industry. In the sale of potatoes and potato products, more emphasis will be placed on quality and environmental aspects. Transgenic potatoes will only be accepted by consumers if they hold evident qualitative improvements that are beneficial for all.

GENETICALLY MODIFIED ORGANISIMS (GMOS) IN CROP PRODUCTION: METHODOLOGIES AND MONITORING

The Food and Agriculture Organization (FAO) hosted an Expert Consultation on "Genetically Modified Organisms in Crop Production and Their Effects on the Environment: Methodologies for Monitoring and the Way Ahead" from 18 to 20 January, 2005 in Rome.

The main objective of the consultation was to review the scientific basis for, and procedures to establish, effective post-release monitoring of genetically modified (GM) crops and develop guidelines to strengthen member countries' capacities to design and carry out monitoring programmes. The participants represented a wide range of expertise from research institutes, universities, international agencies, regulatory agencies, the private sector and civil society. The consultation was jointly organized by the Plant Production and Protection Division of FAO's Agriculture Department and the Inter-Departmental Working Groups on Biodiversity for Food and Agriculture and on Biotechnology in Food and Agriculture.

The experts emphasised that GM crop deployment must comprise the whole technology development process, from pre-release risk assessment to biosafety considerations and monitoring post release. The positive and negative effects of GM crops on the environment are shaped by location and context, and monitoring programmes should recognize that there are important sources of variation within and among farming systems. It should inform decision making and provide feed back to the regulatory process and policies that support the development of sustainable practices. Wherever possible, the objectives of monitoring programmes should, therefore, be nested within processes that address broader goals.

The experts did not list or evaluate individual indicators needed for monitoring, but emphasized the critical importance of planning the process. Major outputs of the meeting were:

i) A review of scientific criteria and procedures that address the technical aspects of monitoring environmental effects of GM crops;

ii) Two strategies that could be used as the basis of efficient monitoring programmes and,

iii) Recommendations for scientists managing the monitoring process, policy and decision makers, FAO and other relevant international agencies.

The capacity to undertake monitoring varies globally. Several developed countries have undertaken large-scale, long-term research and post-release monitoring programmes for GM crops that have provided an effective basis for decision making. Monitoring programme development is, however, a greater challenge in the developing world, where possible hazards are less clearly understood and the stakeholders are less well defined. In addition, opportunities for engagement in public debate are limited, environmental protection measures are less effectively enforced, and there are insufficient resources for research and development or for strengthening local expertise.

To address these challenges, experts have developed a robust design for monitoring that could work within limited resource levels, using the example of herbicide-tolerant rice in Asia with the potential risk of gene flow to weedy rice. The core values of the monitoring programme are the serious commitment to engage and consult with people with a stake in the final outcome, and a judicious selection of indicators that meet the basic requirements for scientific rigour and address stakeholder concerns, and can trigger appropriate management or regulatory responses.

The key steps or actions for developing a monitoring programme are as follows:

- Set monitoring programme goals and immediate objectives
    - Consult stakeholders, including farmers and managers, regarding the natural resources to develop the goals and immediate objective.
- Identify potential barriers
    - Prioritize and develop plans to overcome or minimize potential field barriers or otherwise.
- Identify potential risks and benefits
    - Use stakeholder and expert knowledge of potential risks/concerns and benefits of GM crops, and ways and indicators to measure these factors.
- Develop a testing hypothesis to guide actions and decisions
    - Ensure that the hypothesis is simple, robust and can be easily tested in the field.
- Identify a limited number of potential indicators
    - Ensure that the indicators meet the basic requirements of scientific rigor;
    - Reflect key elements of the hypothesis tested;
    - Compare with control sites and/or baseline values prior to GM crop release; and
    - Estimate the status and trends in indicator values.
- Determine appropriate trigger values for decision making and action
    - Anticipate the range of decisions and actions if triggers are exceeded; and
    - Prepare a follow-up action plan.

- Cultivate a transparent and effective process
  - Ensure follow-through continued involvement of stakeholder;
  - Maintain clarity in analysis and reporting, and identify needs; and
  - Build linkages with policy development and capacity building.

The consultation viewed these actions as occupying a toolbox. They should not be adopted as an inflexible, linear process. Full stakeholder engagement should be fostered through formal and informal networks, alliances and initiatives to promote resource mobilization, communication and information dissemination. Building trust and transparency is the only way to sustain an effective link between monitoring and the resulting actions.

The Food and Agriculture Organization (FAO) hosted an Expert Consultation on "Genetically Modified Organisms in Crop Production and Their Effects on the Environment: Methodologies for Monitoring and the Way Ahead" from 18 to 20 January in Rome. The main objective of the consultation was to review the scientific basis for, and procedures to establish, effective post-release monitoring of genetically modified (GM) crops and develop guidelines to strengthen member countries' capacities to design and carry out monitoring programmes. The consultation was a follow-up to the earlier FAO Expert Consultation on "Environmental Effects of Genetically Modified Crops"1 which had recommended that the environmental effects of GM crops be assessed on a case-by-case basis and emphasized the emerging need to monitor possible medium- to long-term environmental impacts through adequate practical methodologies.

The meeting was a three-day event organized by the Plant Production and Protection Division (AGP) of FAO's Agriculture Department. It was co-sponsored by the FAO Inter-departmental Working Groups (IDWG) on Biodiversity for Food and Agriculture and on Biotechnology in Food and Agriculture. Seventeen experts from around the world were invited to participate in their personal capacity, including representation from the scientific community, international research centres, private sector and the civil society. A background paper on monitoring was prepared and distributed to all participants.2 The consultation was inaugurated by Louise O. Fresco, Assistant Director General of FAO's Agriculture Department. She welcomed the participants and emphasized the Organization's commitment to providing tools to assist countries in making their own informed choices on the matter, as well as protect the productivity and ecological integrity of farming systems. She urged the experts to consider the importance of networks and partnerships for practicability and cost-effectiveness, and to provide access to necessary information and enable its dissemination, should nations introduce post-release monitoring to address both foreseen and unforeseen impacts of GM crop production. She felt confident that FAO would be better positioned to assist member countries in making appropriate choices in this area from the recommendations received from the broad range of expertise assembled in this meeting.

The Director of AGP, Mahmoud Solh, stressed the need for evaluating current monitoring methodologies and procedures, identifying the common elements and constraints so that the Organization can provide guidance for strengthening member countries' capacities to establish effective monitoring of GM crops, as appropriate. He emphasized the facilitator role of FAO in the development of a follow-up mechanism for monitoring medium- to long-term environmental effects of GM crop cultivation involving United Nations agencies, Consultative Group on International Agricultural Research (CGIAR) centres and other international and national centres.

Peter Kenmore, Chairperson of the IDWG on Biodiversity for Food and Agriculture, introduced the Provisional Agenda which was adopted unanimously. He briefly described the process proposed for the consultation. The first section would be devoted to presentations on current monitoring procedures, country experiences, large-scale experiments on monitoring GM crops and management of monitoring programmes. This would be followed by the two thematic group discussions where the experts would analyze proposals from the perspective of (a) countries with well-established risk assessment procedures and scientific infrastructure, and (b) countries that have more limited capacities. He emphasized that the scope of the consultation was post-release monitoring and hoped that effective guidelines and recommendations would be developed through the deliberations.

Thereafter, the chairperson of the sessions invited the speakers to present their papers, followed by general discussions. On the final day, the meeting was closed with the adoption of the preliminary meeting report and with the draft recommendations.

The experts considered that it was important to properly define monitoring and to outline the role of monitoring in relation to other environmental data collection and analysis procedures. Monitoring was defined as a procedure that involves the systematic measurement of selected variables and processes that may be affected by a given practice. Reasons for monitoring include the need to meet environmental protection goals, concerns about deviations in ecological integrity from a predetermined standard or verification of risk assessment findings.

Monitoring does not substitute for rigorous risk assessment in protecting against adverse environmental impacts, although unlike risk assessment, it may also be used to quantify the potential benefits of GM crops.

Successful monitoring procedures build upon existing ecological data sources that establish the status of the system under investigation. Monitoring should not be confused with general environmental surveillance or ecological inventory: monitoring is goal-oriented, and designed to detect change in comparison to reference sites, and/or pre-treatment condition. When effective, monitoring addresses the priorities of people with a stake in its outcome, and feeds back to inform management and policy development.

Deployment of GM crops must encompass the whole process of technology development from pre-release risk assessment to post-release monitoring. Monitoring programmes should recognize and take into account important sources of variation between farming systems and GM crop types in order to properly address potential interactions between the GM crop and the environment. The positive and negative effects of GM crops will vary with location and context, and monitoring will require a new model of working in order to inform actions at the farming system level. The capacity to undertake monitoring varies globally and reflects the level of ecological knowledge associated with particular systems, the local capacity to plan, implement and analyze the data, and the integrity of the pathway that leads from the data to decision making, and back to effective management.

Paul Jepson reviewed monitoring principles based upon the Expert Consultation background review paper. Analysis of long-term biodiversity monitoring in agro-ecosystems tends to be retrospective, with time lags between data collection, analysis and response. Monitoring of functional, often abiotic, indicators has a better record for early detection of adverse impacts.

Decision making and effective responses are only possible when plausible mechanisms underlying effects are known, and when monitoring analysis has high inferential power.

Measurements must also translate to the values and concerns of stakeholders in the final outcome if management responses are to be implemented. Post-release monitoring must consider functional, taxon-based and structural indicators to detect the drivers of change associated with GM cropping. Some farming systems will be more sensitive than others.

Sensitive systems may be at intensification limits or ecologically fragile, with high species turnover rates and poor connectivity with natural areas. They may also be critically dependent upon the grower knowledge base, R&D support may be poor, and the policy environment may be inflexible.

Angelika Hilbeck discussed monitoring biodiversity and ecological functions in the context of European Union Directive 2001/183 which requires monitoring for all GM commercial releases. Monitoring designs must be case-specific (to verify risks) and general (to detect unanticipated effects). A project of the German Federal Agency for Nature Conservation is identifying faunistic indicators using a species-ranking approach, which characterizes and ranks species by ecological function, occurrence, spatio-temporal abundance and relevance, and an impact pathway approach, which identifies hazard scenarios using 'event-tree analysis' and 'fault-tree analysis'. The two tools are used in succession; the first prioritizes species based on characteristics and conservation goals independent of genetically modified organisms (GMOs) and the second subjects them to fault- and event-tree analyses to identify species at risk.

Janice Thies discussed methodologies for monitoring the soil ecosystem and its function. The soil provides many ecosystem services including decomposition and nutrient cycling. The agricultural soil food web, with crop residues as its base, includes decomposers (bacteria and fungi), and predatory protozoa, nematodes and micro-arthropods. GM crop residues have the potential to disrupt energy and material flows, and monitoring should be designed to detect detrimental changes in trophic structure and/or key ecosystem services. Soil scientists are yet to agree upon the factors that determine soil ecosystem integrity, and the level of change that might trigger concern. Promising indicators include the level of retention and form of soil organic matters4, soil respiration rate, abundance of shredder species (collembola and mites), microbial biomass, nitrogen mineralization and nitrification, soil glomalin concentration, and molecular indices of soil community structure.

- Before/after comparisons, or comparisons with control (without GM crop) areas, are essential if analysis of monitoring data is to have inferential power. Data must span the whole cropping system.
- Background data required for all systems includes soil parameters, climatic conditions and crop management (fertilisers, crop protection chemicals, crop rotations and previous crop history).
- Existing biodiversity measurements and abiotic measures of system conditions should be collated,

and availability of monitoring expertise must also be established.

- Monitoring should focus on potential positive and adverse effects of concern to stakeholders.
- Available data on the turnover of GM crop residues in the soil should be compiled into a global database.
- Scientific experiments, undertaken by researchers to develop understanding of mechanisms, do not constitute monitoring; they are, however, essential precursors to effective monitoring because they provide a direct link between measures of change and the mechanisms that underlie such change if it is occurring.

Detlef Bartsch discussed the impact of monitoring GM crops on the environment. GM crop environmental risk assessment in the European Union (EU) identifies areas of uncertainty, including the potential for large-scale and long-term cumulative impacts that should be addressed by monitoring. The types of variables to be monitored must be identified with the procedures to measure them and an appropriate time period for measurement. Monitoring designs must be within logistic limits. Monitoring can also be linked with conservation goals, e.g., via the EU Directive on environmental liability. Damage in this context can include effects on aquatic and terrestrial protected areas and natural habitats, with reference to a baseline or conservation status, ecosystem services that are offered, and the capacity to recover. Damage is not considered to have taken place if impacts consist of fluctuations within normal variability, effects of natural events or normal management, short-term effects, or improvements in condition. Agro-ecosystems may already be included in national environmental monitoring programmes, and surveillance systems may already exist. Having a legal definition of damage may help to focus the monitoring effort and make it more costeffective.

Leslie Firbank discussed the farm-scale evaluations (FSEs) of spring-grown GM crops in the United Kingdom (UK). They constituted a very large experimental regime, and were not designed as monitoring studies. Biodiversity impacts of genetically modified herbicidetolerant (GMHT) sugar beet, maize, spring oilseed rape and winter oilseed rape were evaluated in separate experiments, each with 60–70 replicates that represented UK farming environments. Herbicide regimes in GMHT sugar beet and spring oilseed rape reduced weed numbers more than conventional crops, with effects on invertebrates. Currently, these two crops are not allowed to be grown in the EU. Weed numbers were higher in GMHT maize and commercial growing was allowed. The requirements for ongoing monitoring should be based on an understanding of what is an unacceptable impact on biodiversity. The same results in a different part of the world may give different policy responses if the conservation goals differ or if the balance between environmental, social and economic goals differs.

Fleur François provided a regulatory perspective on approaches and challenges in conducting risk assessment and monitoring in New Zealand, which has regulated GMOs since the late 1980s. Over 50 GMO field tests have been conducted but no GMOs have been released. The Hazardous Substances and New Organisms (HSNO) Act 1996 requires consideration of the sustainability of native and valued introduced flora and fauna, intrinsic value of ecosystems, public health, Mori (indigenous people) culture and traditions, economic costs and benefits and international obligations. Applications to release GM crops are declined if they fail to meet minimum standards relating to environmental impact. Monitoring may be required for conditional release approvals, if technically feasible and cost-effective. Post-release 12 monitoring of GM crops is not considered a substitute for adequate pre-release risk assessment.

Robert Blackshaw outlined approaches to studying the environmental effects of GM crops in Canada, where 5 million hectares of GM crops are grown annually. A 12-year field experiment is examining environmental and economic effects of herbicide-tolerant (HT) canola, maize and potato (until 2003), and *Bacillus thuringenesis* (*Bt*) maize. Data include soil quality and weed seed bank at initiation, weed density by species (species shifts), assessments of resistance development, target insects and plant diseases, arthropod community dynamics (diversity), soil microbial biomass and diversity, transgenic DNA persistence in soil, *Bt* toxicity persistence in soil, DNA transfer to soil microorganisms, crop yield and quality, and production economics. A second study addresses an HT canola seed in the soil seed bank. It was pointed out that although much scientific evaluation is conducted before GM crops are approved for commercial production, post-commercialization studies are prudent because some environmental impacts of GM crops are likely to be scale- and/or time-dependent.

Bao-Rong Lu outlined methodologies for monitoring environmental effects of GM crops in China, with special emphasis on rice. Biosafety research has been funded on GM cotton, rice, soybean, wheat, tomato and *Brassica* species, including gene flow and its ecological consequences, impact of transgenes on non-target organisms, changes in biodiversity, development of *Bt* resistance, fitness of inter-specific hybrids, and field performance of GM crops. Research on rice and its wild relatives provides a model for selfing, wind-pollinated crops. It addresses pollen flow, crop-to-crop and crop-to-wild gene flow, biodiversity influences of GM rice, fitness performance of hybrids between GM rice and wild rice species, and cost-benefit analysis. The objectives are to determine the most effective methodologies for monitoring environmental effects of GM crops and to develop guidelines for safe management.

Eliana Fontes presented details of monitoring for the environmental effects of GM crops in Brazil, where agricultural crops are grown in all five geographical regions, which differ in topography, climate, ecological and socio-economic characteristics and biodiversity. New agricultural technologies must fit within a culturally diverse society, a mega-biodiverse country and subsistence to industrial farming systems. Field trials of GM crops have been held since 1997, but only GMHT soybean is commercially cultivated. There are concerns about adverse effects on non-target organisms, and some crops have sexually compatible wild, feral and backyard relatives. Gene flow may pose a threat to the long-term preservation of crop species' genetic diversity. The diversity of agricultural systems in Brazil and the variety of expertise and baseline information needed for monitoring, poses a significant challenge. A Post-Commercial Monitoring Plan required by the National Technical Biosafety Commission for commercial release of GMHT soybean and an impact assessment of *Bt* cotton were presented.

Gurling Bothma discussed field experience and methodologies for monitoring the environmental effects of GM crops in South Africa, where GM yellow and white maize, soybean and cotton are grown. Monitoring by seed suppliers is required by the Office of the Registrar: Genetically Modified Organisms Act, 19975 to ensure refugia are maintained. Seed 13 companies have established a GM Seed Standing Committee to coordinate an Insect Resistance Management system and a protocol is under development. Indirect monitoring of seed sales is also used to monitor the maintenance of refugia in cotton, to prevent resistance build up. A different strategy is used to manage and monitor compliance by less technologically advanced farmers. Companies selling the GMHT crops are required to monitor for herbicide resistance in weeds, but this has not been detected yet. Monitoring and management systems are being synchronized in South Africa to make them accessible across the diverse farming community.

- The types of variables to be monitored must be identified with the procedures to measure them and appropriate time-periods for measurement. Monitoring designs must be within logistic limits.
- All biodiversity effects in the UK FSEs arose from the effects of herbicides whose use was enabled by the GM technology rather than because of the mode of crop breeding.
- Although the FSEs were not monitoring studies, their design criteria (i.e., procedures built from a clear hazard scenario with an identified mechanism) were equivalent to those required in monitoring programme design.
- Several countries already have good procedures in place that provide a useful model for implementation in other countries, and some countries have made a commitment to conduct long-term research on monitoring environmental effects of GM crops.
- Several countries that have adopted GM crops do not have a monitoring process in place yet. In rice, to date, the major concern has been the presence of wild rice relatives and the impacts of the foreign gene in these species. The level of out-crossing between transgenic cultivated rice and weedy rice is still low; however, it may change as the infestation

increases. Procedures are also needed to monitor the impact of GM soybean and cotton, but in several cases, countries did not have trained personnel or resources allocated to this purpose.

- In one example, the private sector has shown interest in investing in the monitoring process, but there is not enough human capacity to carry it out.
- Policy makers vary in their capacity to exploit details about GM crop ecological effects, and ecological impact data vary in the degree to which they can inform and assist the development of effective policy. Emphasis in some policy arenas tends to be on crop production goals, whereas in others (e.g., the EU), ecological effects are a priority.
- Monitoring must consider factors of concern to stakeholders, and to be effective, they must establish a relevant location, scale and duration. The specific GM traits may guide design, as may significant changes in crop management.
- Capacity building for GM crop monitoring is needed in developing countries. There should be a responsible institution/organization in the country to coordinate monitoring.

CGIAR centres may help with regional implementation and play a role in information gathering.

Richard Visser reviewed GM potato work in the Netherlands and Peru in collaboration with Maria Scurrah. Monitoring of GM potato for volunteer plants in the Netherlands has occurred since 1990. For GM crops in centres of origin, special additional procedures are required, including analysis of gene flow, investigations of pollinators and pollen flow. These procedures were developed in GM nematode-tolerant potato6. In the high Andes, improved varieties of *Solanum tuberosum* spp. *andigena* mix with the seven other cultivated and wild species. Gene flow was quantified, with overlapping flowering periods, sexual compatibility, presence of pollinators, and seed survival. Hybridization between cultivated and wild species occurred despite chromosome and endosperm balance differences, and more hybrids were obtained than predicted.

Rodomiro Ortiz presented experience with monitoring GM crops in the International Maize and Wheat Improvement Center (CIMMYT); one of the CGIAR centres. Its goal is to improve low diversity traits and generate public-sector-provided products, which include drought-tolerant wheat and insect-resistant maize. A public-awareness campaign includes food, feed and environmental safety, monitoring of resistance and establishment of refugia, non-target effects and gene flow.

Monitoring of genetic resources is a CGIAR-wide concern, with emphasis on the quality of genebanks. Decisions, policies and procedures about monitoring should be science-based, and this requires education, an area where CIMMYT/CGIAR can play a role. There will be a need to continue to evaluate the need for, and type of, monitoring as new (and unique) products are developed and released.

Coosje Hoogendoorn discussed the adventitious presence of transgenes in CGIAR *ex situ* collections. A 2004 workshop provided genebank managers with measures to adopt in response to requests for GM-free material7. High-risk crops currently include maize, which is wind cross-pollinated and has a sexually compatible wild species, Teosinte, in Mexico and Central America. Varieties may be protected by applying isolation distances and rotation.

There is a need to develop screening tools and to ensure that best practices are adopted. Other high-risk crops, now or in the future, include canola, sorghum, pigeon pea, millet, Cruciferae, sunflower and forage grasses.

Raymond Layton provided an industry perspective of monitoring strategies and management of GM crops. Monitoring should be designed to test a hypothesis and it should be conducted only if recommended by scientifically based risk assessment. Monitoring studies should be located and designed to reduce uncertainties. The controls and end points should be clearly defined before monitoring is conducted. Important questions to be answered prior to 15
monitoring include: "What are we seeking to protect?" and "How will the data be used?" Trained personnel and appropriate sample collection and analysis techniques are needed to ensure that the data will be useful. The audience for monitoring must be clearly defined and personnel who interpret and communicate results should be trained.

Suman Sahai discussed development of socio-economic indicators to assess the impact of GM crops. Socio-economic impacts of GM crops are relevant in a developing country context where livelihoods could be affected. Indicators for GMHT crops include changes in family income due to wage loss and shortage of weeding impact on health and veterinary care (loss of medicinal plants), impact on household nutrition and family income (loss of fodder for livestock and loss of supplementary crops grown on field bunds and field margins), soil erosion through loss of vegetation cover, and development of HT-tolerant weeds and the costs of eradicating them.

The impact of using *Bt* crops should be assessed by monitoring the impact on lepidopteran resistance development that may be caused due to overuse of *Bt* transgenes. Measurements of the impact on organic agriculture, crop diversification, mixed farming and inter-cropping are needed, as well as agro-ecosystem and adjoining natural ecosystem effects, and the impacts on traditional farming practices and indigenous knowledge.

- GM crop monitoring is an international issue. The CGIAR centres, relevant UN agencies, national and international centres and universities should assist in the development of effective procedures.
- The experts recommended that the biotechnology industry should work with the public sector. The majority of the information collected by the industry is not in the public domain, and a greater degree of sharing is needed. The capacity to do risk assessment and monitoring is often lacking in developing countries.
- Socio-economic indicators may also need to be developed to address monitoring of GM crops, especially in the context of developing countries.
- Raising public awareness and building confidence among all stakeholders is essential for establishing a successful monitoring programme.

The experts were asked to focus on the scientific criteria and procedures for effective protocol design and to broadly address the technical aspects of monitoring. Some of the conclusions reached are itemized below. The experts were unanimous in concluding that monitoring programmes need to be developed in ways that recognize important sources of variation between farming systems and GM crop types. The effects (both positive and negative) of GM crops will vary with location and context, and monitoring will require a new model of working in order to inform actions at the farming system level. The main conclusions reached in the working sessions are summarized below.

The experts discussed data needs and development of

minimum datasets. The challenge will be to address variation within and between countries in: (i) regulatory requirements; (ii) the organisms, process and systems to be monitored; and (iii) individual goals for monitoring programmes.

All possible sources of data should be taken into account and identified including biodiversity surveys and inventories, soil databases, genebanks, plant protection services, farmer 16 organizations, private sector (including sales figures), plant variety rights agencies, prerelease monitoring databases, environmental groups and water authorities.

The experts recommended that coordinators of post-release monitoring be appointed (possibly from the lead GM regulatory agency) for coordinating the collection of data, compiling the information in an appropriate way and performing the analysis and reporting. The challenge will be to link data sources and systems that were not set up for this purpose.

The experts made a case for the broad surveillance of practices in farming system that are to include GM crops. The specifics of the monitoring programme depend on the GM trait, and the farming system and the broader (natural and managed) habitat context. Agricultural systems have unique social, economic and environmental properties.

The experts also presented several challenges for the scientific and technical development of monitoring including differences between farmers, environmental groups and agencies in perceptions of risks and benefits, lack of available expertise, absence of extension services and lack of available resources.

Two separate expert working groups undertook programme design exercises, using examples that reflect the range of capacities to develop and undertake monitoring.

They proposed processes and mechanisms for developing a monitoring programme that meet the needs of country or region with a) substantial knowledge of potential hazards and programmes for monitoring environmental effects of GM crops and b) limited knowledge of potential hazards and little experience in monitoring environmental effects of GM crops.

- Developed countries have the infrastructure to undertake monitoring, but there is no consensus on the types of questions to be addressed or basic data requirements
- Stakeholders can be polarized, with broader society concerned about adverse effects, adopting farmers focused on positive effects and non-adopters (e.g. organic farmers) concerned about adverse impacts on livelihood
- The experts were optimistic that monitoring could work, within reasonable resource levels.
- The outline for programme design was considered to be a powerful basis for developing a monitoring system.
- The monitoring system will work best if nested within other processes that address wider goals, otherwise the process can easily become burdened with multiple tiers of questions and concerns.
- Stakeholder engagement is intrinsic to the system, from the beginning right through to the end. It is vital to build trust, legitimacy and transparency. It is the only way to deliver an effective link between goals on the one hand and triggers and decisions on the other.
- Expertise is available in both the formal and informal sectors, but it needs to be identified and engaged.

- Collaborate with UNEP to build capacity with the National Biosafety Framework, and the Biosafety Clearing House.
- Establish pilot workshop processes on a small scale in several areas to work the process through as a thought experiment and establish pilot systems that include collection, management and reporting of field data.

In their discussions under Thematic Session 3 (Sharing experiences, international context and Networking) the experts developed a series of recommendations and follow-up actions to be carried out through sharing experiences and networking. It was agreed that a monitoring programme should incorporate existing environmental surveillance and ecological inventory data, and the available expertise in monitoring and taxonomy. They must also consider the organisms, functions, ecological and socio-economic processes that stakeholders value, and would seek to have protected. Post-release monitoring can work, even within the restricted resource levels, but only if there was a continuous engagement of all the stakeholders. This has to be fostered through formal and informal networks, alliances and initiatives which promote communication and information dissemination. The outcome of the monitoring programme must inform decision making. It should feed back the regulatory processes and policies that support the development of sustainable agricultural practices. The experts agreed to a monitoring system that would be implemented on a case by case basis and nested within broader environmental goals. It was more important to get imperfect monitoring systems up and running quickly, in circumstances where these are required, rather than wait until we have perfect systems.

In this context, the experts discussed the role and

contribution of the international community in the process of establishing effective monitoring procedures, including UN agencies, CGIAR centres, and national and regional centres of excellence. The FAO and other international organizations have a major responsibility to start a process to develop a comprehensive understanding of country and local community needs with respect to postrelease monitoring of GM crops. Their recommendations are as follows:

Scientists and International Community Managing Monitoring Programmes

- The scientific community is strongly encouraged to engage in research, development and education associated with the effective implementation of post-release monitoring programmes. Critical and innovative thinking was essential to develop new and appropriate methodologies.
- Identify and mobilize relevant expertise, especially field and traditional expertise, as well from biotechnologists, biologists, ecologists and environmental scientists. Include expertise from other fields, like social sciences. Engage scientific societies.
- Involve stakeholders early and continuously in the process.
- Collaborate and develop inventory(ies) and biodiversity assessment in agro-ecosystems and neighbouring natural habitats, to provide baseline data and current trends coupled with measurements of agricultural practices and the patterns and distribution of crops that can assist in determining potential indicators.
- Participate in data sharing mechanisms including access via the Internet, where appropriate.

- Avoid selection of inappropriate indicators by following a robust process: o Define the amount of change in any recommended indicator that should trigger concern and what aspects of the environment and cropping/soil management practice that might affect (increase or decrease) trigger values o Gain awareness of all potentially useful datasets, and identify the most robust (precise/accurate) sources of existing data (regionally, nationally, internationally) that might be used as the indicator or as a surrogate.
  - o Define the most relevant scale, time-frame(s), at which the indicator operates to guide sampling and analysis.
  - o Ensure that appropriate, accessible methods exist to measure recommended indicators with the precision required.
- Improve dialogue between stakeholders and scientists by focusing stakeholder input towards specific questions you wish to address. The process should be transparent, comprehensive and include an education and information dissemination programme for stakeholders.

Policy and Decision Makers at the Regional and National Level

- Identify clear goals and specific objectives for environmental monitoring programmes, and when/where these programmes are appropriate. To achieve this engage stakeholders to the greatest extent possible to understand what your society values and what their main interests and concerns are for deployment of GM crops. Competing policy goals exist and should be integrated

- Carefully identify the values (e.g. environmental, cultural, and economic) to be protected to analyse whether implementing a monitoring programme would protect those values or allay concerns?
- Responsibility for monitoring and reporting are national, but programmes can be undertaken using sub-national levels or jointly among countries.
- Ask definitive questions. Formulate a monitoring programme to measure effects that are connected with clearly stated protection values. State the amount of change over a defined time scale in any recommended indicator that should trigger concern. This requires setting thresholds and quantifying effects, including defining statistical detection limits.
- The process should be transparent, comprehensive and include an education and information dissemination programme for stakeholders.
- Develop policies to involve and strengthen public institutions, and to build capacity to develop, maintain and learn from well constructed monitoring programmes. Priority must be given to educational programmes and capacity building for relevant stakeholders (farmers, consumers, public, etc.)
- Identify what actions need to be taken in response to information from a monitoring programme. If it is unclear for what purpose monitoring data will be used, the monitoring programme will be ineffective and irrelevant. Additionally outcomes of the monitoring programme should inform public debate.
- Determine trigger criteria and action plans for intervention, remedial action.
- Ensure that any requirements set forth are feasible in terms of costs, personnel, expertise, protocols,

and relevance of data generated. Adequate resources are required for monitoring programmes. Funding may sourced through partnerships between the public sector, biotechnology industry and other private sectors, and various stakeholder groups.

FAO, CGIAR Centres and International Organizations

• The FAO has a big responsibility to initiate the process and continue the dialogue started among stakeholders with respect to monitoring.

- Build upon the process to develop a comprehensive understanding of country needs and local communities. Be prepared to take on a stewardship role as the need arises.
- Support establishment of Pilot Monitoring Projects for collection, management and reporting field data as appropriate through joint initiatives.
- In countries/regions where CGIAR centres are located, they should provide national/regional support. For crops under their mandate, they should provide global support and serve as repository of regional information that has been deemed of sufficient quality that "mining" for monitoring change can occur. Provide the expertise to use those data for, regional meta-analyses. In some cases the centre will be the source of the GM technology and it will have special responsibilities to insure that independent, rigorous monitoring procedures are established.
- FAO, UNEP and other international and regional organizations collaborate to build national capacity for monitoring programmes, facilitate data management, leverage funding, partnerships and collaborations for monitoring programmes.

GENETICALLY MODIFIED LACTIC ACID BACTERIA: A CASE STUDY

Ever since human domestication took place approximately 10,000 years ago, parts of our food supply have been cultured in order to support survival of the human population. In our current times, the main objectives of agriculture are still the same, the assurance of efficient food supply in the first place, and the increase of quality of the supplied foods for the consumer in the second place. Modern biotechnology, using recombinant DNA technology, offers ample opportunities to further improve food production in the expanding world of today. Within the United States of America there is not much public resistance towards the use of genetically modified organisms, GMOs, in food. However, in other areas of the world, for instance Oceania and Europe, the production of novel foods using GMOs is subject to strong debate.

For example, within the European Union the scientific community states that there is sufficient legislation to guarantee that such novel foods are safe for the consumer and the environment, but green groups in many cases state an opposite opinion. The support for the use of GM crops and foods among the European population has recently increased in some countries. Yet at the same time a majority of Europeans do not support agri-food biotechnology (Gaskell et al. 2003). In the current debate about the application and safety of GMOs in the food industry no clear distinction is made between GMOs derived from animal, plant, or microbial origin. Moreover, ethical aspects could play a role in the debate about GMOs from animal or plant origins, but these are considered less relevant in the discussion about genetically modified micro-organisms. Apart from this, the current European legislation on novel foods and the use of GMOs in food production processes, as laid down in the regulation EC No 258/97, amended in regulation

1829/2003, directive 2001/18/EC (revised directive 90/220/EC), directive 90/219/EC, and Commission Recommendation 97/618/EC, generates a framework for what should and what should not be permitted by law. The legislation, however, is not yet completely clear on a number of important scientific matters. First of all, the legislation predominantly focuses on the methodology rather than on the end product and hence, holds too strongly to the definition of GMOs. Consequently, organisms in which the genetic material has been altered by recombinant DNA techniques in a way that does occur naturally, for instance by point mutations or small deletions, are considered to be GMOs. Secondly, foods with a new structure which date from before 1997 are not considered as novel foods. Thirdly, self cloning (Box 2) of non-pathogenic micro-organisms is not considered to lead to a GMO as long as containment of the organism is guaranteed (directive 90/219/EC).

However, the same self-cloned micro-organism when used in products (deliberate release) is considered to be a GMO (directive 2001/18/EC). Moreover, the market introduction of foods with organisms that have been improved via classical breeding or random mutagenesis is currently acceptable without profound safety analysis[1], but it could be questioned whether these foods are more safe than GMO-derived novel foods. The new regulation 1829/2003 (amending 258/97) regulates the approval and the labelling, and distinguishes between products from GMO, and products which are produced with help of GMO. Foods, ingredients, additives, flavours as well as feedstuffs, which contained or are GMO, or which were produced from GMO are in the scope of the regulations. The labelling of these products is also process-oriented. Products produced with help of GMO are not in the scope of the regulations. It is still unclear if the regulations can be applied on the use of GMO (micro-organisms) within fermentation (Jany and Schuh, 2005). The rationale for food safety regulations is the assurance to prevent any adverse effects on human health or on the environment. The starting point for a food safety approach is the comparison between the new foods and the conventional counterparts that have a history of safe use, both for the food produced and for the micro-organism that is modified. This approach leads to identification of new or altered hazards relative to the conventional counterpart (substantial equivalence, Box 3) is essential (OECD, 1993).

This document will analyse the current situation within Europe with regard to the safe use of novel foods derived from genetically modified lactic acid bacteria (GM-LAB). The application and safety assessment of genetically modified yeast and fungi are not discussed; neither is the use of GMOs derived from plants (Kuiper et al. 2001) or animals. Genetically modified microbial inoculants for use in agriculture, such as biological control agents, biofertilisers, or phytostimulators is reviewed elsewhere (Amarge, 2002; Morrissey et al. 2002). An overview will be given of the specific genetic techniques that generate GM-LAB and the advantages that the novel foods fermented with GM-LAB may bring to the consumer or the producer. Moreover, the current safety assessment procedures are considered and further suggestions are made for an adequate risk analysis when GM-LAB are applied in food production. Careful analysis of the properties of GM-LAB combined with a profound safety assessment procedure for foods that are produced through the actions of GM-LAB, are essential for comparing the safety of such products with traditional foods. Finally, based upon the principle of assuring food safety for the consumer, it could be argued to make no distinction between foods produced by LAB that have been altered in an uncontrolled way, for instance by random mutagenesis, or a more directed way, such as GM-LAB. This could imply that all foods fermented by LAB should be subjected to a safety assessment procedure, before commercialisation can occur. Recently, such issues have been considered for non-GM-LAB in a draft document introducing the concept of Qualified Presumption of Safety (QPS), which will be discussed below (EU, 2003). Furthermore, we will examine the value of the principle of a long and safe history of use of traditionally used LAB and discuss whether this could be a starting point for the acceptance of GM-LAB.

Lactic Acid Bacteria and Genetic Engineering

LAB have a long history of use by man for food production and food preservation. LAB are Gram-positive, non-spore forming bacteria and naturally present in raw food material and in the human gastro-intestinal tract. The heterogeneous group of LAB includes the rod-shaped bacteria like lactobacilli, and cocci such as streptococci, lactococci, pediococci and leuconostocs. LAB are widely used as starter cultures for fermentation in the dairy, meat and other food industries. Their properties have been used to manufacture products like cheese, yoghurts, fermented milk products, beverages, sausages, and olives. These food-grade bacteria can also improve the safety, shelf life, nutritional value, flavour and quality of the product. Moreover, LAB can be used as cell factories for the production of food additives and aroma compounds. It is further assumed that LAB may function as probiotics and contribute to the general health of the consumer upon consumption. The use of probiotics falls currently within a grey area between food and medicine and many health claims assigned to probiotics are not yet scientifically proven.Another application - the use of LAB in the production of proteins for application in health care or for development of new vaccines (Mercenier et al. 2000) – is more related to pharma than to food. In the future it is predicted that knowledge about the interaction between LAB and the human host will open new avenues for developing LAB which support human health. The uncontrolled genetic alterations of LAB that occur during random mutagenesis may lead to strains with improved traits. These may be either attractive for the manufacturer of fermented foods, or have benefits for the consumer. Both directed and uncontrolled genetic alterations result in a change of the genetic code of the micro-organism that may affect the transcription and translation processes and, consequently, may influence metabolic processes in the cell. However, it seems that in the current legislation regarding the use of GMOs in food industry, the nature of the DNA modification could dictate the necessity for the novel food product to be subjected to a safety assessment procedure.

Uncontrolled Genetically Altered LAB

Spontaneous mutations may occur in LAB by natural events such as insertion sequence elements (Visser et al. 2004), radiation, erroneous DNA replication or transcription, and other factors. The level of such mutations depends upon the growth conditions. By screening of natural isolates of LAB, strains with improved fermentation characteristics can be selected. The frequency of mutations can be further increased by exposing LAB to mutagenic conditions such as UV light or chemicals like N-methyl-N'-nitro-N-nitrosoguanidine (NNG), or ethyl methyl sulphonate. Specific screening for desired traits may result in the identification of new strains with improved utility for the food industry.

Examples of LAB Selected After Uncontrolled Genetic Alterations

Selection of strains that have been subjected to uncontrolled genetic alterations is used in the dairy industry to improve certain intrinsic characteristics of the fermented end product, like flavour, structure, nutritional value, or phage resistance. Spontaneous mutations in single genes can lead to altered lactose metabolism, citrate uptake, and increased proteolytic activity. In *Lactobacillus bulgaricus* a spontaneous insertion sequence (IS) element-mediated deletion of the *lacZ* gene altered lactose metabolism resulting in limited fermentation capacity. As a consequence, yoghurt, made with these altered *L. bulgaricus*, is not suffering from post-fermentation acidification during the shelf life period (Mollet and Delley, 1990).

A gene inactivation method for several lactic acid bacteria has been described that is mediated via IS elements that may integrate into a gene, resulting in inhibition of function (Maguin et al. 1996). According to current legislation, however, strains with inactivated genes obtained by this latter IS-based strategy are considered to be GMOs. *L. lactis* strains that are involved in the fermentation of Roquefort cheeses were randomly mutated, and selection was based on carbon dioxide production that improves the quality of the cheese (El Attar et al. 2000). In a similar way LAB defective in lactate production, but with increased production levels of the butter flavour compound diacetyl were selected after random mutagenesis with the mutagen NNG (Boumerdassi et al. 1997). The selection of a spontaneously mutated *L. lactis* strain that overproduced diacetyl, responsible for the butter flavour in many fresh dairy products, is described by Monnet et al. 2000. A stronger and more direct selection procedure was applied by the temporarily introduction of a plasmid that generates valine prototrophy in the bacterium (Curic et al. 1999). After selection of the desired mutation in the *aldB* gene, encoding a-acetolactate decarboxylase, the plasmid was removed, leading to strain that did not contain any foreign DNA. Nevertheless, under the actual EU legislation this strain is still considered as a GMO after release in the environment (Henriksen et al. 1999). This last case is a perfect example of possible inconsistencies in the current legislation. The nutritional value of fermented foods can also be improved by (over)production of healthy components, like vitamins, in the fermenting bacteria. To increase vitamin B2 (riboflavin) production levels in the dairy starter bacterium *L. lactis*, bacterial cells were exposed to increasing concentrations of roseoflavin. This resulted in the selection of a roseoflavin-resistant strain with deregulated riboflavin biosynthesis and increased production levels of riboflavin, up to 1 mg/L.

Food products obtained from fermentation with this modified *L. lactis* will contribute strongly to the daily recommended intake (DRI) of 1.3 mg for riboflavin. Partial analysis of the genome of this mutant revealed a mutation in a supposed regulatory sequence just upstream of the first gene of the riboflavin gene cluster (Burgess et al. 2003). LAB can also be selected for the removal of undesirable compounds from raw food materials. In traditional yoghurt fermentation, the undesirable sugar lactose is degraded only partially. One of the degradation products is galactose which may be harmful for people suffering from galactosemia and which also may cause cataract problems (Liu et al. 1998). Vaughan describes the selection of spontaneous galactose-fermenting mutants of *Streptococcus thermophilus*, that contain up-mutations in the *gal* operon and that may assist in removal of undesired galactose from the food matrix (Vaughan et al. 2001). The selection of a double mutant strain of *L. lactis* with an inactivated glucose uptake system deficient in glucokinase could also reduce the concentrations of lactose and galactose during fermentation. Moreover, this strain could also be used as natural sweetener due to the excretion of the undegradable lactose moiety glucose (Thompson et al. 1985).

Controlled Genetic Alteration of LAB

An alternative for random mutagenesis is targeted or directed modification of the DNA. Renault (2002) has evaluated the potential benefit to society and the possible risks associated with the use of genetically modified lactic acid bacteria in food and health applications. It was concluded that risk assessment and expected benefits will determine the future use of modified bacteria in the domains of food technology and health. Controlled genetic modifications vary from single base pair substitutions, mutations, insertions of genes into the chromosomes, or removal of DNA from the chromosome (deletions) resulting in inactivation of specific enzymes.

According to the current legislation, all such directed genetic alterations lead to strains that are considered as GMOs, except for strains obtained by self-cloning, transduction and conjugation. A variety of techniques has been developed to generate GM-LAB, such as cloning systems, chromosome modification systems and expression systems (de Vos, 1999). The most popular transformation system for generating directed genetic alterations in LAB is electroporation with self-replicating vectors. Alternative systems are conjugation and transduction (Gasson, 1990). Besides efficient cloning systems, adequate expression systems have been developed allowing the controlled expression of homologous and heterologous genes. Controlled constitutive expression is possible by using a system of synthetic promoters (Solem and Jensen, 2002), whereas a nisin induced controlled expression system (NICE) allows the gradual over expression of genes (Kleerebezem et al. 1997). Other systems are controlled by promoters based on sugar utilisation, e.g. the lactose operon promoter (Payne et al. 1996), by presence of salt, *e.g. gadC* promoter (Sanders et al. 1997), pH decrease (Madsen et al. 1999), or temperature up-shifts (Nauta et al. 1997) (reviewed by de Vos, 1999). Targeted gene replacement or removal and inactivation of genes can also be applied via (non-replicating) vectors using the natural event of crossing-over during cell division and DNA replication (Leenhouts et al. 1996). Compared to the use of replicating vectors in GM-LAB that result in new or enhanced cellular behaviour, the deletion of genes after double cross-over events (by using a non replicating plasmid) does not result in the addition of any DNA to the genetic content of the cell.

A specific aspect related to the application of vectors in industrial strain improvement is the use of selection markers. The use of antibiotic resistance markers might result in transfer of antibiotic resistance from one organism to another. As a consequence, the practical value of antibiotics that are used for human health or in veterinary practice will be severely reduced. Hence, food-grade resistance markers are preferred. Currently there are many food-grade selection markers for vector cloning in LAB. For instance, transfer of the á-galactosidase gene (*aga*) and a gene coding for a putative transcriptional regulator from the LacI/GalR family (*galR*) of *Lactococcus raffinolactis* ATCC 43920 into *L. lactis* and *Pediococcus acidilactici* strains modifies the sugar fermentation profile from melibiose negative (Mel(-)) to melibiose positive (Mel(+)) (Boucher et al. 2002). A similar food-grade vector is based on complementation of the lactose operon in *L. lactis* NZ3600 or *L. casei* by introduction of *lacF*, or *lacG,* respectively, enabling growth on lactose (MacCormick et al. 1995; Platteeuw et al. 1996, Takala et al. 2003). An alternative system is based on a suppressor tRNA allowing growth in milk of a purine auxotrophic strain (Dickely et al. 1995). A newly developed food-grade marker is characterised by the requirement of D-alanine in the medium to enable growth of the micro-organisms (Bron et al. 2004). An overview of general strategies for constructing food-grade markers has previously been reported (de Vos, 1999). As mentioned before, the current use of these food-grade markers if constructed via self-cloning techniques is restricted to applications with contained use. Other targeted modifications of the genetic content of DNA may occur via conjugation and transduction (Gasson, 1990). These processes are considered natural events. According to the current legislation, bacteria that are changed by using these transfer systems are not considered as GMOs.

Examples of GM-LAB Engineered Via Directed Genetic Alterations

Directed mutagenesis is widely applied in research to improve fermented food products. The following paragraphs will provide an overview of successful examples of metabolic engineering aimed at the improvement of certain characteristics of food products, like flavour, structure, shelf life, nutritional value, or product performance in general. A schematic overview of GM-LAB. The flavour and flavour stability of buttermilk was improved by inactivation of the *aldB* gene encoding á-acetolactate decarboxylase (Swindell et al. 1996). The over expression of peptidase genes in LAB via self-replicating plasmids could also enhance flavour formation and cheese ripening. Peptidolytic activity can be increased by overproduction of peptide degrading enzymes originating from the same species or from other species. An example of the self-cloning strategy is the overproduction of an aminopeptidase (PepN) by *L. lactis* (Gasson and de Vos, 1994). GM-LAB with increased proteolytic propertied are generated by the over expression of the genes encoding PepN, PepX, PepC and PepI peptidases from a highly proteolytic *L. helveticus* strain (Joutsjoki et al. 2002) or PepI, PepL, PepW, and PepG from *L. delbrueckii* (Wegmann et al. 1999) into *L. lactis*. Moreover, the expression of *gdh* from *Peptoniphilus asaccharolyticus*, encoding glutamate dehydrogenase, into *L. lactis* increases the production of á-ketoglutarate (Rijnen et al. 2000). This enhances the degradation of amino acids, which also benefits the cheese ripening process. Food fermentation processes that, because of size, cannot be operated under strict axenic conditions, may suffer from bacteriophage infections resulting in lysis of starter bacteria. The transformation of industrially important strains with phage resistance genes from other LAB could generate new phage resistant strains. Engineering of starter strains aimed at inhibiting phage development is possible by creating a system of lethal gene induction upon phage infection (Djordjevic et al. 1997) or by production of antisense RNA inhibiting translation of phage RNA (Kim et al. 1992; Walker and Klaenhammer, 1998). The inactivation of a phage infection protein, *pip*, involved in phage adsorption and DNA injection resulted also in a phage resistant *L. lactis* strain (Monteville et al. 1994).

The lactococcal abortive infection mechanisms AbiA and AbiG were introduced into *Streptococcus thermophilus*, and a range of phages capable of infecting this host were sensitive to these mechanisms (Tangney and Fitzgerald, 2002). The quality of food fermentation processes could be further improved when they are protected from spoilage by other bacteria. The expression of the food-grade lantibiotics in dairy starter strains could prevent spoilage from Gram positive bacteria during the process of cheese ripening (Horn et al. 1999). The production of lantibioticsto prevent growth of *Listeria monocytogenes* and *Lactobacillus fermentum,* during the production of cheese was also achieved via the non-GMO approach of conjugation (Hugenholtz and de Veer, 1991; O'Sullivan et al 2003). Bacteriophage-resistant dairy starter cultures were also selected after the natural event of transduction (Hill et al. 1989; Gasson, 1990) or after conjugation (stacking) of plasmids with phage resistant genes (Mills et al. 2002).

The event of phage induced cell lysis is undesirable during early stage of cheese making, but cell lysis in a later stage of the fermentation can actually decrease ripening time and enhance flavour due to the release of many enzymes. The controlled production of lysin and holin by bacteria has created the possibility to induce cell lysis at desired states of food fermentation (de Ruyter et al. 1997). Another important feature of food products is texture. Complete gene clusters, encoding exopolysaccharide producing enzymes have been transformed from one LAB strain to another one. The newly generated strains could influence viscosity and texture of the fermented product (Germond et al. 2001). In *S. thermophilus* the phosphoglucomutase gene was inactivated resulting in improved exopolysaccharide production enhancing the viscosity of the fermented food product (Levander et al. 2002). Engineering of exopolysaccharide production in *L. lactis* was also achieved by using a self-cloning strategy.Genetic engineering using genes from non related micro-organisms could also be used to produce high added value products, such as l-alanine. By introduction of a *Bacillus subtilis* (formerly *B. natto*) alanine dehydrogenase gene into a *L. lactis* strain deficient in lactate production, pyruvate conversion was pushed in the direction of alanine. The subsequent inactivation of the host gene encoding alanine racemase leads to the production of the stereo specific and thus highly valuable l-alanine (Hols et al. 1999). The nutritional value of fermented foods can be improved via use of GM-LAB. It was shown that by overexpressing the complete riboflavin gene cluster (four genes) via a multi copy plasmid and strong promoter, riboflavin production reached levels as high as 8 mg/L (Burgess et al. 2003). With this production level in fermented dairy products, average servings would already contain the DRI of riboflavin. Another important vitamin in the human diet is folate. This B-vitamin decreases the risks of neural tube defects in newborns and lowers the concentration of homocysteine in the blood, which is a risk factor for the occurrence of cardiovascular diseases. Nowadays, a large part of the population in both developed and under-developed countries does not meet the DRI of folate (200-400 μg). By the overexpression of all the genes encoding for the folate biosynthesis pathway in *L. lactis*, folate levels ranging from 5 - 8 mg/L can be achieved (Sybesma et al. 2003a). This could result in fermented food products that require an intake of less than 100 ml to meet the DRI for folate. It should be stated here that both riboflavin and folate-overproducing strains were developed as proof of principles, and in the current state contain plasmids with chloramphenicol resistant markers. These strains could only be applied in food fermentations after the substitution of the antibiotic resistance marker with a food-grade marker. Another line of research focuses on control of the bioavailability of folate produced in fermented foods. In nature, folate predominantly exists as polyglutamyl folate, however folate is absorbed as monoglutamyl folate. Several studies show that the bioavailability of monoglutamyl folate, which does not need enzymatic deconjugation, is higher than that of polyglutamyl folate. The overexpression of the first enzyme in the folate biosynthesis pathway increases the production of monoglutamyl folate (Sybesma et al. 2003a; Sybesma et al. 2003b).

Recently, the engineering of a *L. lactis* strain was described that contributes to the formation of lactose-free and naturally sweetened foods. By the deletion of three genes involved in glucose uptake and degradation, and introduction of the genes for lactose-PTS and tagatose-6P pathway, the strain can grow on lactose. The galactose-moiety of lactose is used as a C-source, whereas the glucose moiety is completely excreted (Pool et al. 2003). This GM-LAB obtained by self-cloning has identical properties to the earlier selected non-GM-LAB of Thompson (Thompson et al. 1985). However, the genetic background of the latter strain was spontaneously altered and, except for the desired mutations, it is likely that several non-anticipated mutations have occurred (de Vos and Hugenholtz, 2004). An example of the use of recombinant DNA technology in the development of probiotic strains is in removal of undesirable sugars from the gastro-intestinal tract. á-Galactosides such as raffinose and stachyose, that are present in raw agromaterials like soy, cause digestion problems in a lot of people. Because humans do not produce intestinal á-galactosidases, these sugars pass into the lower GI-tract where they are fermented by gas-producing bacteria, rendering intestinal pain and flatulence.

The expression of genes encoding enzymes that are able to degrade raffinose, stachyose, etc, such as á-galactosidases (Boucher et al. 2002; Silvestroni et al. 2002), in certain LAB, results in the (extracellular) production of sugar-degrading enzymes. Consequently, upon passage of these sugars through the GI tract, they will be degraded by the probiotic LAB and not by the resident gas-forming bacteria (Piard, 2003). In the feed industry, strain improvement can also be successfully applied. *Lactobacillus plantarum* is used as an inoculum in grass silage and vegetable products due to its pH lowering effect, which allows preservation of the products (Seale, 1986; Daeschel et al. 1987). The cloning of the phytase gene, *phyC*, from *Bacillus subtilis* in *Lb. plantarum* has recently been described (Kerovuo and Tynkkynen, 2000). Phytases are enzymes which hydrolyse phytate. These enzymes have a significant value as feed additives. The application of a phytase-producing silage strain in feed preparation may improve the nutritional value of feed and, at the same time, reduce the environmental phosphate pollution following consumption by monogastric production animals. Another GM-LAB designed to improve the nutritional value of feed is an amylolytic *L. plantarum* silage strain with improved starch degrading capacity (Fitzsimons et al. 1994).

Risk Factors for Use of Genetically Modifed Lactic Acid Bacteria

The process of introducing novel traits into LAB by the addition, substitution, removal, or rearrangement of defined DNA sequences, including the DNA sequences used for the maintenance or transfer of the new DNA into the recipient strain could lead to the generation of LAB with novel genetic properties and a modified cellular behaviour. It is anticipated that this will result in the intended effects. However, unintended effects, that could be a consequence of the modification, could also occur in the newly generated LAB. A distinction can be made between the predictable and unexpected effects that differ in their origin. Predictable effects are a foreseeable and direct consequence of the intended genetic alterations and unexpected effects are caused by unintended genetic alterations, or they may occur as a consequence of predictable effects.

The Predictable Unintended Effects

The predictable unintended effects could occur as a consequence of directed genetic alterations. For example, by insertion or modification of genes that are located in an operon, the expression of adjacent genes that are present in the same operon may be affected (polar effects). Also the expression of the inserted gene and subsequent production of a protein may result in potential toxicity or allergenicity of this protein, especially when posttranslational modifications in the recipient micro-organism may be different from the original host of the new protein. Another kind of predictable unintended effect may arise as a result of the synthesis of a chimeric protein (fusion protein), due to the insertion of foreign DNA. These new proteins may provoke a different type of cellular behaviour, but they might also lead to toxicity or allergic responses by the consumer because of their different chemical structure. Careful analysis of the genomic organization prior to and following the directed genetic alterations could be used to avoid the occurrence of such predictable unintended effects. For instance, the use of well-characterised loci for the insertion of the new DNA may prevent the effects on expression levels of adjacent genes. It should be noted that despite the fact that strains obtained via random mutagenesis are not considered as GMOs, the occurrence of such predictable unintended effects is very likely to occur within such mutant strains as well. In addition, it is possible that DNA elsewhere in the genome will also be affected by random mutagenesis. This could result in synthesis of modified enzymes and altered metabolite production (unexpected unintended effects), which might affect the health of the consumer. Therefore, it is not unlikely that certain GM-LAB may have a lower risk profile than LAB isolated after random mutagenesis.

The Unexpected Unintended Effects

The occurrence of unexpected and unintended effects could be a consequence of unintended genetic alterations. However, these effects could also be caused by the direct intended effect or by the predictable unintended effects of the alterations. The overproduction of certain desired metabolites or blocking pathways preventing the production of specific metabolites in the cell may result in unexpected and undesired cellular responses. For instance, a LAB defective in lactate dehydrogenase capacity may produce other acids and alcohol under certain environmental conditions. Increased concentrations of desired metabolites may also result in feedback inhibition of parallel biosynthetic pathways in the cell. Hence, a careful analysis of alterations in metabolism is desirable. Recently it was shown that inactivation of *ldhD* (responsible for the formation of D-lactate) in *Lactobacillus johnsonii*, primary and secondary end products were not easily redirected in high amounts to other pathways (Lapierre et al. 1999). Another unexpected and unintended effect may occur when the overproduction of one product inhibits the production of another compound. For instance, the biosynthesis of the vitamins folate and riboflavin starts with GTP as a substrate. An increased flux through the riboflavin biosynthesis pathway, as found with *L. lactis* producing increased riboflavin levels (Burgess et al. 2003), might reduce the concentration of GTP available for folate biosynthesis.

As a consequence, folate production levels may be decreased. Another class of unexpected unintended effects is the change of the microbial population when GM-LAB are applied in mixed cultures. Several starter strains may grow simultaneously or in succession. For instance, during yoghurt fermentation, *S. thermophilus* grows prior to *L. bulgaricus* which only starts growing after acidification of the milk by the *S. thermophilus*. During the fermentation of sauerkraut, heterofermentative bacteria such as *Leuconostoc mesenteroides* initiate the fermentation, and after 3 to 7 days it is usually succeeded by the more acid-tolerant homofermentative *Lactobacillus* species, while *Lactobacillus plantarum* completes the fermentation (Lu et al. 2003). Hence, the synthesis of new metabolites, specifically those with antimicrobial activity, may disturb the balance in a mixed culture resulting in unintended changes in the final product. The phenomenon of gene transfer can probably be seen as one of the most important unexpected unintended effects. In the previous section many examples are given about self-replicating vectors to introduce additional DNA in the LAB cell. The production and consumption of GM-LAB will result in their contact with micro-organisms in the environment, including the intestinal microbiota of the consumer. Consequently, transfer of DNA between these micro-organisms may occur (horizontal gene transfer) (Jain et al. 2002; Doolittle et al. 2003). Transfer between the GM-LAB and the consumer has not been demonstrated (Genereux and Logsdon, 2003). Horizontal gene transfer is not restricted to plasmid-based DNA, as other modes of gene transfer such as those mediated by transposons, phages, or naked DNA, can also take place (Droge et al. 1998). In case a selective advantage is conferred by transferred DNA, the likelihood of stable gene transfer may be higher.

For this reason the use oftransferable antibiotic resistance markers in fermenting micro-organisms should not be allowed, because these could provide a selective advantage following transfer to intestinal micro-organisms as soon as consumers are treated with antibiotics. This is a realistic scenario since there is evidence indicating that gene transfer of antibiotic resistance genes has occurred to bacteria that are present in traditionally fermented foods (Teuber et al. 1999). In contrast, the inactivation of genes by deletion does not result in the addition, but in the omission of DNA. Consequently, any discussion about potential gene transfer in GM-LAB with inactivated genes is void. It should be noted that the potential for the occurrence of unintended effects is not restricted to the use of recombinant DNA techniques. Rather, it is an inherent and general phenomenon that can also occur in the development of strains using traditional genetic techniques and procedures, or from exposure of micro-organisms to the intentional or unintended selected pressure. An overview of these concerns and other potential effects of food-related genetically modified micro-organisms is provided by von Wright and Bruce (2003).

Considerations for Restrictions in Development GM-Lab

Partially following earlier recommendation (Wymer, 1998), and to ensure the safety of humans and the environment, the following restrictions regarding the uncontained use ofGM-LAB are recommended: 1. Only LAB with a qualified presumption of safety (see below, EU, 2003) should be used as a recipient; 2. The use of transferable antibiotic resistance markers should not be permitted; 3. The expression of distantly related genes in LAB is acceptable under contained use only, unless the inserted gene will not induce growth or survival advantages to other micro-organisms after potential horizontal gene transfer; 4.The strain engineering should never result in production of toxic or allergic compounds or generation of a pathogenic strain that may affect human health.

Food-Grade Micro-Organisms

The use of micro-organisms in food production is accepted when they have a long history of safe use. However it is not scientifically defined what is a long history and what is a safe use. Moreover, a recent draft document (EU, 2003) suggests that only micro-organisms that have a long and safe history of use in food and that also have a qualified presumption of safety status (see below) should be applied in the food industry. These strains could be used as recipients for genetic modification. The use of pathogenic micro-organisms should not be allowed. However, the history of safe use in food should not be an everlasting guarantee that the strain could always be applied in food fermentation processes. If new research shows that strains with a long history of safe use are producing toxic components in levels that may harm human health, these strains should not be accepted anymore for use in food fermentations. For instance, certain LAB have an unblemished history of safety in food fermentation, but, as was discovered later, may produce unfavourable amines under some conditions (Ten Brink et al. 1990; Cantoni et al. 1994; Faeth and Radler, 1994). Evidently these strains should not be used in preparation of foods without a profound safety assessment, that, especially in this case, also investigates the actual concentrations of the harmful compound to which the consumer will be exposed.

Transferable Antibiotic Resistance Markers

The use of transferable antibiotic resistance markers in LAB used in food production may increase the risk of their transfer to human intestinal flora, and cannot be allowed. The use of safe and sustainable food-grade selection markers is recommended (de Vos, 1999). However, contrarily to current legislation, the use of antibiotic resistance markers in intermediate strains that are used for the generation of new bacterial strains with improved characteristics, and that are completely removed before the final production strain is developed, should not be rejected on the basis of reasons related to antibiotic resistance. For instance, the selection of the earlier described GM-LAB with inactivated á-acetolactate decarboxylase (Curic et al. 1999) should be allowed, because the vector containing the antibiotic resistance gene was removed from the selected strain. It is noteworthy that antibiotic resistance could also occur spontaneously. Well known examples include resistance to streptomycin and rifampicin that are a consequence of point mutations in rRNA or RNA polymerase genes, respectively. A specific case is resistance to vancomycin. Strains carrying transferable vancomycin resistance genes should not be used for food fermentations. Related Gram-positive bacteria like enterococci can spontaneously acquire vancomycin resistance. In such vancomycin resistant strains, spontaneous mutations in genes encoding cell wall synthesizing enzymes may result in altered peptidoglycan intermediates bearing D-alanine–D-lactate termini that confer vancomycin resistance (Arthur et al. 1996; Baptista et al. 1997). Similarly, LAB that produce D-lactic acid show endogenous resistance to vancomycin. These naturally occurring phenomena however, are not a reason to exclude such strains from consideration as recipients in constructing GM-LAB, since this endogenous resistance is not transmissible. Specific consideration needs to be given to tolerance to antibiotics that is caused by certain classes of transporters. Many of these are naturally present in LAB and may be essential for their survival in certain environments, where they could be involved in export of toxic compounds. It is conceivable that these transporters also export structurally related compounds, such as some antibiotics. This is likely to be the case for the endogenous tetracycline resistance found in many lactobacilli. In most cases the level of antibiotic resistance is far from that found in clinically relevant antibiotic-resistant pathogens. Hence, there appears to be no need for specific precautions when such antibiotic tolerant LAB strains are used as recipients. It should be further noted that the eventual selective pressure on these transporters caused by antibiotics used during medical treatment is not as dominant as the selective pressure caused by all the (natural) compounds that exist in nature. Therefore, it is likely that the risk of antibiotic resistance transfer via genes coding for transporter proteins from harmless bacteria to clinically important bacteria is much higher in nature than in the situation where fermenting bacteria are consumed.

Expression of Homologous or Heterologous Genes

The potential event of horizontal gene transfer following expression of foreign genes in LAB used in food fermentations is one of the major concerns that affect the safety assessment for consumer and environment. When the applied genetic elements originate from LAB that have a long and safe history of use in food, it is often said that no new risks are introduced. Although this argumentation is very logical, it does not mean that fermented foods are by definition safe. Therefore, adequate safety assessment remains necessary. However, it could eliminate the risk of negative consequences for the environment. For microbiota in complex natural habitats it is generally assumed that all genetic information is already omni-present, but that natural constraints drive selection. Thus, fear of negative consequences after the release of engineered LAB (containing genetic elements of other LAB species in the environment is unfounded. Evidently this is also valid for strains that have been subjected to spontaneous DNA modifications, or deletions of DNA. The introduction of more distantly related genes in micro-organisms is likely to occur in nature at much lower frequencies compared to the transfer of related genes. The cloning of such genes on a plasmid can be regarded as a potential risk factor since horizontal gene transfer could lead to selective advantage in specific niches. Therefore, some precautions are suggested for the expression of heterologous genes in LAB from non-microbial origin. As long as heterologous genes are only inserted in the chromosome via double cross over, and the inserted heterologous gene does not lead to evolutionary advantages and survival of such strains, there are no reasons to fear the consequences of potential gene transfer. Containment of such engineered strains may prevent interactions with the environment. In addition, survival in the environment can be limited by using host strains with a strict dependency on growth factors or special temperature or pH -requirements (Steidler et al. 2003). Genes that may cause pathogenicity (*e.g.* generation of virulence factors) or that encode for enzymes that are known to be involved in synthesis of toxic or allergic compounds, or precursors of toxic or allergic compounds, should *a priori* be excluded as targets for genetic engineering. Finally, one should consider avoiding the use of sequences that mediate integration into other genomes in the development of genetic constructs. Although in nature such sequences widely exist, and their use in strain development could be regarded as safe, it is better not to accelerate and support the process of gene transfer by the use of these sequences.

Considerations of the Safey Assessment Procedures for Foods Devided from Traditional Lab and GM-LAB

With the exception of those LAB not previously used in the preparation of a human food within the EU, LAB for food use are not subject to EU regulation. Implicit in this absence of any formal requirement for a safety assessment is the recognition that there has been a long history of presumed safe use. The long and safe history of LAB usage is in itself an arbitrary criterion and may need further consideration. If the long history of safe use is based on the absence of reports of the occurrence of adverse effects on consumers of fermented foods, it can be questioned whether this could also apply to certain GM-LAB. This would include LAB obtained by self cloning, or LAB that were subjected during their production process to recombinant DNA techniques, but that did not maintain the foreign DNA. This would imply the immediate acceptance of all such non-pathogenic GM-LAB for food production. However, is this whole debate not made for the assurance of safety for the society? Before regulation about food safety was an issue, society decided autonomously that it wanted to consume certain fermented foods, probably based on positive experience. These facts could be a convincing social, but less scientific, argument that LAB with a long history of safe use may be accepted in food fermentations until adverse effects are reported. The European Food and Feed Culture Association (EFFCA) has produced a list of starter species for which a documented history for use in food manufacture exists. Documented history of use is defined by the EFFCA as cultures sold for human consumption in quantities exceeding 100 kg of freeze dried culture. Although for many species belonging to LAB the safety documentation is sufficient, it is remarkable that the list also contains species that are potentially pathogenic.

Qualified Presumption of Safety

Within the EU it is proposed to introduce a system similar in concept and purpose to the GRAS (Generally Recognized As Safe) definition used in the USA, which could be applied to micro-organisms and their products. Such a system should lead to a listing of qualified micro-organisms that will not have properties that may adversely affect human health or the environment (EU, 2003). For this purpose, Qualified Presumption of Safety (QPS) is being introduced as an assumption based on reasonable evidence. It aims to provide a qualified generic approval for non-genetically modified micro-organisms, without requiring all organisms used in food production with a long history of safe use to be subjected to a full and unnecessary safety review. A case-by-case safety assessment then could be limited to only those aspects that are relevant to the organism in question. A pre-requisite for QPS would be that identity is unambiguously established at the taxonomic level claimed. Appropriate and state of the art biochemical and molecular biological methods must be applied to realise this. For organisms not commonly used in food production or without a long history of safe use, this implies the need for experimental data on the genetics of the taxonomic unit and the growth and biochemical characteristics of the component strain under a variety of relevant environmental conditions. For some groups of organisms, such as those used as biocontrol agents for plant protection, a consideration of the impact on the wider environment may be appropriate. However, this could exclude organisms considered either to be of uncompromised origin and regularly introduced into the wider environment or originating from soil or water. In both cases the organisms are naturally occurring and therefore free of any need for an environmental impact assessment.

The QPS could only be valid for bacteria that enter the food chain and that are free of any acquired resistance to antibiotics of importance in clinical and veterinary medicine. The presence of antibiotic resistance determinants would not exclude their safe use for production purposes provided that only the fermentation products are retained in the final product. A decision scheme relating to the concept of QPS has been proposed and is presently subject to discussion (Figure 1) (EU, 2003). It is worthwhile mentioning that this QPS decision protocol might be subjected to modifications before its implementation. The decision step questioning other concerns related to safe use of micro-organisms, which are not involved with pathogenicity or toxicity, is currently not well defined. Many industrial strains will be the product of mutagenesis and selection programs designed to improve their phenotype for a particular purpose. These cryptic mutations will not affect taxonomic status. As a consequence, following the QPS decision tree, strains that are not produced via genetic modification, but contain altered DNA sequences that may cause predictable or unexpected unintended effects, will be accepted for use in food preparation without a further safety assessment. The next paragraph considers whether such strains of LAB should be subject to similar scrutiny as GM-LAB.

Due to the long history of use of many LAB, the resultant fermentation products are obviously not systematically assessed for safety. Moreover, food products containing improved strains obtained after random mutagenesis are not assessed either. It could be reasoned that the omission of adequate safety assessment in the past does not render the fermented foods safe for the consumer and the environment in the future. However, in the debate about safety of food derived from genetically modified microorganisms, the point is not to demonstrate that novel foods are completely risk-free, but that the risks are comparable to those associated with traditional foods. The safety assessments that have been conducted thus far are primarily aimed at discrete chemical entities such as food additives, or specific chemical or microbial contaminants that pose identifiable hazards and risks. It is likely that unintended effects could always occur in the development of strains using traditional (non-recombinant DNA) techniques, or from exposure of micro-organisms to selective pressure or certain environmental conditions, or just by natural events. This would justify that each new natural LAB strain should undergo the same safety assessment procedure as engineered strains. Whether this is a viable approach is currently being investigated by analysing differences at the global transcriptional level in strains isolated by random mutagenesis and by directed mutagenesis (Renault, 2001). The development of GM-LAB for use in food should be focused on minimizing the occurrence of unintended effects with adverse consequences on human health. Risk assessment studies using new technologies should focus on global response in cellular metabolism as a tool to identify unintended effects. It could be considered that in the future the decision to grant a QPS status to strains that are isolated after spontaneous or induced random mutagenesis, or to assess those strains using a profound safety assessment procedure as obliged for novel foods, would be taken after the performance of representative comparative profiling studies. These may include transcriptomics, proteomics, and metabolomics, between the mentioned strains and their natural conventional counterparts. In case of no relevant differences between the strains, the QPS protocol could be applied. In analogy with the outcome of these studies, the profiling of engineered strains obtained by directed genetic alterations may result in a more scientific food safety assessment procedure for novel foods derived from these GM-LAB, in which not the technical aspects, but the final products are assessed[2].

Consideration of Safety Assessment Procedures of GM-LAB and Novel Foods

The safety assessment of food produced by GM-LAB should comprise analysis of the intended effect, the nature of the modification, and detectable unintended effects that may occur in the micro-organism or in the food that is producedby the fermenting micro-organism. Furthermore, issues that are specific to the use of recombinant GM-LAB include the genetic stability[3], the potential for gene transfer (see above), the colonisation of the gastrointestinal tract and the persistence therein, interactions that the recombinant-DNA micro-organism may have with the gastrointestinal flora or the mammalian host, and any impact of the recombinant-DNA micro-organism on the immune system. The development of modern genetic techniques could enhance the safety assessment, and the follow up by post market surveillance may also contribute to the safety assessment and increase the consumers' trust. Furthermore, the safety assessment procedure should encompass the obligations stated in directive 2001/18/EC and Commission Recommendation 97/618/EC.

Toxicity and Allergenicity

Because the nucleic acid techniques enable the introduction of DNA that can result in the synthesis of new substances in the LAB, specific attention should be paid to the toxicity and allergenicity of the newly synthesised compounds (proteins and metabolic products). When a substance is new to foods or food processing, the use of conventional toxicology studies or other applicable studies on the new substance will be necessary. In the case of proteins, the assessment of potential toxicity and allergenicity should focus on amino acid sequence similarity between the new protein and the known protein and anti-nutrient activities (*e.g.* protease inhibitors, lectins) as well as stability to heat or processing and to degradation in appropriate representative gastric and intestinal model systems. Appropriate oral toxicity studies may need to be carried out in cases where the protein present in the food is not similar to proteins that have been previously consumed safely in food, taking into account the biological function of the protein in the donor organism where known.

At present there is no definitive test that can be relied upon to predict allergic response in humans to a newly expressed protein. However, it could first be assessed whether a newly expressed protein is one to which certain individuals may already be sensitive as well as whether a protein new to the food supply is likely to induce allergic reactions in some individuals as could be predicted by amino acid sequence homology and sera screenings. The Codex Alimentarius (2003) proposes that IgE-cross-reactivity between the newly expressed protein and a known allergen should be considered a possibility when there is more than 35% identity in a segment of 80 or more amino acids. The development of international serum banks would enhance efficient testing for allergic potential of the newly expressed proteins. Strategies such as stepwise contiguous identical amino acid segment searches may also be performed for identifying sequences that may represent linear epitopes. The size of the contiguous amino acid search should be based on a scientifically justified rationale in order to minimise the potential for false negative or false positive results. Moreover, resistance to pepsin degradation has been observed in several food allergens and was found to correlate with the allergic potential of the protein under scrutiny. Therefore, the resistance of a protein to degradation in the presence of pepsin under appropriate conditions indicates that further analysis should be conducted to determine the likelihood of the newly expressed protein being allergic. Potential toxicity of substances that have not been safely consumed in food before should be assessed case-by-case on the basis of the identity, concentration, and biological function of the substance and dietary exposure. The type of studies to be performed may include evaluations of metabolism, toxicokinetics, chronic toxicity/carcinogenicity, impact on reproductive function, and teratogenicity. It is important that in such studies the stability of recombinant DNA in lactic acid bacteria is also considered. It is also important to ascertain the impact of changes on the overall nutrient profile and to verify to what extent the modified nutrients are bioavailable and remain stable during processing and storage. Recombinant LAB that remain viable in foods may interact with the immune system in the gastrointestinal tract. Closer examination of these interactions will depend on the types of differences between the recombinant-DNA micro-organism and its conventional counterpart.

Modern DNA Techniques, Genomics, Transcriptomics, Proteomics, and Metabolomics

The rapid development of genetic and biochemical techniques can be applied to increase reliable, fast and profound safety assessment of the fermenting micro-organisms and the derived foods, in accordance with the concept of substantial equivalence. It is preferable that the unintended effects could be foreseen at an early stage so that potential harmful effects could be analysed beforehand. The presence of new proteins or the change in concentrations of certain proteins could point to the altered activity of biochemical pathways that may result in changed metabolite content. The analysis of the complete set of metabolites in the cell could also provide information about potential toxicity and about the general nutritional value. Genome wide analysis of modified strains and of their natural counterparts, for instance by using fingerprint techniques, or by complete sequencing of the genome, can reveal all loci in the genome where mutations have occurred that could result in unintended effects. Transcriptome analysis, by using DNA array or DNA chip technology, could provide information about altered expression patterns of the new micro-organisms compared to the wild-type strain. The observation of altered expression levels of genes not involved in the targeted effect of the genetic engineering could be a starting point for the analysis of the reaction products of the enzymes coded for by the genes of interest. This might be done by studying the literature and by application of bioinformatics, and, if a harmful compound may be produced, subsequent biochemical analysis could be performed to measure the presence of these metabolites. More relevant for food safety assessment is an analysis of the proteome and metabolome of the modified micro-organisms. Proteome analysis, by using 2D-gel electrophoresis, 2D-HPLC or advanced mass spectrometry, provides information on altered protein concentration and content produced by the modified micro-organisms. Likewise the outcome of the transcriptome analysis, the reaction products of the new proteins or increased protein concentrations themselves could be predicted and analysed.

The value of proteomics has already been proven in genetically modified rice that was engineered to produce more storage proteins. SDS-PAGE analysis revealed an unexpected increased level of prolamins (Kubo, 2000). Metabolome analysis, by using a wide range of chromatographic and spectroscopic techniques, could provide information about the presence of new metabolites or altered concentrations of existing metabolites. Increased concentrations of known metabolites could be compared with current acceptable daily intake levels to assure the consumer's safety. Likewise, decreased concentrations of vital components could reduce the nutritional value of a new product. The identification of new proteins, or altered protein levels, new metabolites, or altered metabolite levels, could be a starting point for further toxicological studies. To strengthen the safety assessment, transcriptome, proteome and metabolome analyses under food processing and GI-tract conditions should be performed in case the new engineered strains have potential for survival in the human gut. The outcome of these data analyses will also provide information on the occurrence of the intended and, if present, of the unintended effects. The conclusion of the analysis should be the assurance that neither the intended effects, nor the unintended effects, have adverse effects on human health, because no harmful components or proteins with allergic characteristics are being produced. In the near future it is possible to develop a universal proteome and metabolome reference frame consisting of all proteins and compounds produced by fermenting micro-organisms. The proteome and metabolome reference frame should contain information on potential allergenicity of proteins, based on amino acid sequence comparison with known allergens, and information about acceptable maximum and minimum levels of all compounds present in food. The changed protein and metabolome content of any new engineered strain can then be easily compared with the reference frame. Comparison of the proteome and the metabolome of the new engineered strains with the reference frame could be a starting point for further food safety assessment procedures and could ensure that further safety assessment remains proportional to the perceived risks. For instance, if it could be verified that proteins or metabolites are not produced in harmful concentrations and that the nutritional value is not adversely affected, no further safety assessment would be required for the fermenting micro-organisms and the fermenting food.

Post-market Surveillance

A post-market surveillance (PMS) on novel foods, involves the observation of (health) effects on the consumer over a longer period of time after release of the novel food on the market. It could be questioned whether, after approved pre-market safety assessments of genetically modified foods, the use of post-market surveillance would still be necessary. However, theoretically, novel foods may contain undetected anti-nutrients or toxicants which may lead to (delayed) adverse effects on the health of humans and animals. Moreover, various other factors may also be involved in determining the safety of food in the long run, such as the consumption conditions and the genetic profile of the consumer. Therefore, to further protect the general public, including specific sub-populations such as immuno-compromised individuals, infants and the elderly, and to provide additional information on long term effects, PMS could be applied in the form of a recording and investigation system by national health services, perhaps similar to the adverse drug effects reporting systems used by physicians. Up to now only a few food additives have been subjected to PMS, including the artificial sweetener aspartame (Butchko et al. 1994) and the fat replacer olestra (Slough et al. 2001). Recently within the EU, PMS has started on a novel food related to the use of phytosterol-esters in yellow fat spreads that is assumed to have cholesterol lowering effects in the consumer (Lea, 2002). However, it may be difficult to introduce PMS over a longer time period if the food consumption pattern of individuals is not consistent. It should be taken into account that some consumers might be exposed to newly introduced components from different sources during their daily consumption pattern and consequently might be at risk because of overexposure to certain compounds. PMS could also be used to verify the conclusions about the absence of potential consumer health effects, and it could eventually monitor changes in nutrient intake levels that could occur by the introduction of foods that are likely to significantly change the nutritional status. Although the application of PMS to food products raises significant technical issues, particularly for active surveillance, passive surveillance based on spontaneous reports could be utilised. Apart from this, it could be considered that for novel foods derived from GM-LAB that are substantially equivalent to LAB with a QPS status, PMSwould not be necessary. An effective PMS contains the following elements (Health Council of the Netherlands, 2002): i) Setting up a government supported line for all consumer complaints associated with health and foods. This would enable any side effects produced by a product to be traced, provided that they arise soon after consumption and that they attract attention. One such example is food allergies. This monitoring system will not reveal any causal relationships; however, it can be used to generate a hypothesis. In this way it would alert people to health problems, leading on to a focused follow-up study; ii) Continuous monitoring of consumption of foodstuffs, carried out jointly by government and industry. One precondition is that this must facilitate a detailed breakdown of data on individual products into information on their ingredients (down to the molecular biology level in the case of genetically modified foods); iii) Long term epidemiological, prospective, cohort studies into the relationship between chronic diseases and diet; and iv) An active market monitoring program carried out by companies on novel foods that contain bio-active ingredients. The aim here is to check the accuracy of the presumed (safe) intake of the target group (Health Council of the Netherlands, 2002). It may be concluded that PMS could only prove its value by signalling hazardous events resulting in immediate withdrawal of the concerned food. PMS could not be seen as a safety assessment, but as an early warning system that operates better as the adverse effects are stronger. Moreover, it could be employed if increased confidence by the consumer to use novel foods evolves, and if the principle and reasoning for setting up PMS is well communicated to the consumer.

Suggestions for Cost Accountability of Safety Assessments for Novel Foods

The development, approval and subsequent market release of novel fermented foods involves many economical issues for industry, consumers and society in general. Investments in research and development by the food industry will provide opportunities for employment and the development of new scientific technologies. The consumption of novel foods with added value will benefit the consumer. The introduction of products with proven health claims, may especially contribute to the generation of a healthier community, and this is expected to decrease public health costs. The costs associated in the development of novel foods can only be worthwhile in a society that accepts and buys these novel foods. The introduction of traits that decrease production costs for the manufacturer could be the impetus to reduce the price as a way to generate an additional advantage for the consumer.Conversely, the introduction of traits that benefit the consumer might lead to added value of the product that may justify a price increase to the consumer. Moreover, novel traits could also be developed that have a lower environmental load and that bring advantages to the society as a whole. It is expected that any obligation to perform a profound safety assessment, before the launching of any novel food, could significantly add to the initial costs to be made by industry. The question is, who should cover these costs: the industry, the consumer, national authorities, or supra national authorities? Finally, although each food producer is responsible for the safety of the products placed in the market, in many countries food safety is principally an issue for the regulating authorities. Therefore, it could be proposed that public money could be used to set up the complete infrastructure required for food safety assessment as stated in the paragraphs above, including development of profiling techniques based on transcriptomics, proteomics, and metabolomics. Consequently, it would also be the task of regulatory authorities to control and conclude the whole safety assessment protocol. On the other hand, the industry might be expected to participate as well in the process of assessing novel foods derived from GM-LAB. After approval of the novel food, PMS would remain acombined matter for the society and the industry. The high costs associated with the development and assessment of novel foods may be a limitingfactor for the industry to develop novel foods, especially those with interesting traits based on health benefits for the consumers. Therefore the industry should be allowed to promote novel foods with scientifically proven advantages over traditional foods by use of labelling and other forms of publicity. Currently the possibilities for promoting scientifically proven functional foods are limited and not harmonised within various EU countries. It is evident that the consumer should be protected by prohibiting the labelling of untrue claims. However, not permitting the labelling of true health claims is equally wrong (Katan and de Roos, 2003). The future EU regulation, as proposed in its working paper of 2002 about nutritional and functional claims (EU/SANCO, 2002), seems to provide a good solution for both consumer and industry by allowing and regulating the labelling of health claims.

The Consumer's Preparation

Consumer's perception is an extremely important factor for successful introduction of new technologies in the market. Leisner (2005) described the obstacles for the introduction of pasteurization and starter cultures in the late 19th century to indicate that the gap between industry, consumers and pressure groups is not an entirely new issue. Genetic alteration of plants and animals is sometimes seen as interference with nature that could have unknown and potentially disastrous interactions with human genetics and natural ecosystems. Moreover, many consumers are extremely aware of health threats caused by unknown risks in the food supply. Green groups have been raising several questions about the environmental risks of genetically altered crops and animals. Environmental risks include the evolution of GM-plants into "super weeds," cross-pollination introducing herbicide resistance into existing weeds or introducing undesirable genetic traits into neighbouring crops, and harm to non-target populations caused by toxins introduced to create insect resistance. In the present work we have shown that the consequences of development and use of GM-LAB are different compared to GMO from plant or animal origins. The potential intended and unintended effects and related risks can be predicted more accurately and can be verified. In combination with profound safety assessments, safety for consumers and safety for the environment can be assured. The occurrence of unintended effects is not unique for GM-LAB, but also occurs in LAB used in conventional food fermentations. Logically, this is no reason to ban GM-LAB, but it could be a reason to analyse the unintended effects of traditionally used strains in food fermentation by genomics-based profiling methods. A further response to the public concern is to strengthen the pre-marketing data requirements and to introduce post-launch surveillance program to confirm safety, as described above.

Besides an approval of the safety assessment, the GM-LAB designed for use in food production is probably more acceptable for the consumer if a clear correlation with consumer benefits can be made. Therefore, labelling on food products should emphasise the benefits (eventually health claims) of the novel foods. Evidently, product labels should also inform the consumer about the nature of the (genetically modified) bacteria, the risk assessment procedure, and the nature of the conventional counterpart. However, it is questionable why, according to the new EU legislation, food and feed products and additives that are derived from GM-LAB should be labelled with such a characteristic when there is no DNA or protein derived from a GM-LAB detectable in the final product. A parallel strategy that could be adapted to increase trust among consumers relates to the liability of the manufacturers in case of damage caused by novel foods. Despite profound pre-market safety assessment that is expected to be conducted before the introduction of novel foods onto the market, unpredicted adverse effects directly related to the consumption of the novel food can theoretically not be excluded. The industry and its products would probably gain support if they accept the liability for such events. However, it would not be reasonable to establish an unlimited liability for the industry. Therefore the liability for predictable adverse effects associated with the consumption of novel foods could be placed onto the consumer as long as the labelling includes warnings for the specified adverse effects that could occur in sub-populations. Moreover, the free choice of the consumer should be respected. Thus, alternative similar products, although they will not possess the introduced advantageous traits, should be widely available on the market. A complex and harmonised legislation and regulation system seems at present times the only correct response to consumer demand for protection from the perceived harms that could be caused by GMOs. However, it remains to be seen if the consumers would accept that as a guarantee for the safe consumption of novel foods. For the future it could be recommended to set up governmental sponsored research programs with the ultimate goal of further reducing the uncertainty and mistrust within society. Educating the public on these matters may help to overcome the negative emotions related to the use of certain GMOs.

The low degree of trust by the general public in the (food) industry, in the regulatory authorities and in the (food) scientific community is a major element contributing to the general negative perception of GMOs in the EU. The set up of an infrastructure by the regulatory authorities to subject all fermenting micro-organisms and fermented foods to a new and profound risk assessment procedure might be the extra step that is currently needed to convince all parties involved that novel foods are as safe as traditional foods. A distinction within the discussion and legislation between food safety assessment procedures for fermented foods (dealing especially with LAB), and GMOs derived from plant and animal origin, is necessary in the current controversy. As a starting point, the QPS protocol (EU, 2003), based on taxonomy and on the history of safe use of LAB applied in food, could in the near future be applied to any kind of LAB or GM-LAB provided that a series of modern profiling methods are used to verify the absence of unintended effects by altered LAB that may cause harm to the health of the consumer. Objective safety assessments of GM-LAB could be a way to improve the current legislation. It is expected that knowledge of genome evolution and profiling studies will reveal that in many organisms the genetic material, that is naturally or spontaneously changed, has been modified more than can be established by directed alterations, including genetic modification.

Direct genetic engineering, including self-cloning, generates targeted modifications that can be much better controlled and evaluated than strains that are exposed to random mutagenesis approaches. If such strains are accepted in food fermentations via QPS status, the acceptance by society of GM-LAB obtained via directed genetic alterations could be improved. In line with current legislation, safety issues should dominate the acceptability of GM-LAB in food industry.

A targeted approach should be based upon analysis of key compounds in the new food, such as proteins, carbohydrates, fats, vitamins, minerals, as well as upon analysis of toxicity and allergenicity. A non-targeted approach should use to profile methods to detect global responses to the new food. Global expression profiles will reveal whether changes additional to those expected have occurred. The expression of any new or altered gene in LAB, that is either randomly mutated or self cloned, may affect the transcriptome, proteome, and metabolome profiling and could tell if the expression of the new gene results in any (undesirable) side effects. The assessments of foods derived from modern biotechnology should be reviewed to ensure that the emerging scientific information is incorporated into the risk analysis. In conclusion, broad evaluation of the nature of novel foods derived from GM-LAB, which should be considered as a specific group of GMOs, could be a starting point to bridge the gap between industry, consumers and green groups. This may lead to acceptance of GM-LAB derived novel foods to provide for a better quality of life in today's society.

2

Gene Expression, Molecular Modelling and SNPs

GENE DISCOVERY, EXPRESSION AND ESTS

Investigators are working diligently to sequence and assemble the genomes of various organisms, including the mouse and human, for a number of important reasons. Although important goals of any sequencing project may be to obtain a genomic sequence and identify a complete set of genes, the ultimate goal is to gain an understanding of when, where, and how a gene is turned on, a process commonly referred to as gene expression. Once we begin to understand where and how a gene is expressed under normal circumstances, we can then study what happens in an altered state, such as in disease. To accomplish the latter goal, however, researchers must identify and study the protein, or proteins, coded for by a gene. As one can imagine, finding a gene that codes for a protein, or proteins, is not easy. Traditionally, scientists would start their search by defining a biological problem and developing a strategy for researching the problem. Oftentimes, a search of the scientific literature provided various clues about how to proceed. For example, other laboratories may have published data that established a link between a particular protein and a disease of interest. Researchers would then work to isolate that protein, determine its function, and locate the gene that coded for the protein. Alternatively, scientists could conduct what is referred to as linkage studies to determine the chromosomal location of a particular gene. Once the chromosomal location was determined, scientists would use biochemical methods to isolate the gene and its corresponding protein. Either way, these methods took a great deal of time—years in some cases—and yielded the location and description of only a small percentage of the genes found in the human genome. Now, however, the time required to locate and fully describe a gene is rapidly decreasing, thanks to the development of, and access to, a technology used to generate what are called Expressed Sequence Tags, or ESTs. ESTs provide researchers with a quick and inexpensive route for discovering new genes, for obtaining data on gene expression and regulation, and for constructing genome maps. Today, researchers using ESTs to study the human genome find themselves riding the crest of a wave of scientific discovery the likes of which has never been seen before.

> **An Expressed Sequence Tag is a tiny portion of an entire gene that can be used to help identify unknown genes and to map their positions within a genome.**

ESTs are small pieces of DNA sequence (usually 200 to 500 nucleotides long) that are generated by sequencing either one or both ends of an expressed gene. The idea is to sequence bits of DNA that represent genes expressed in certain cells, tissues, or organs from different organisms and use these "tags" to fish a gene out of a portion of chromosomal DNA by matching base pairs. The challenge associated with identifying genes from genomic sequences varies among organisms and is dependent upon genome size as well as the presence or absence of introns, the intervening DNA sequences interrupting the protein coding sequence of a gene. Gene identification is very difficult in humans, because most of our genome is composed of introns interspersed with a relative few DNA coding sequences, or genes. These genes are expressed as proteins, a complex process composed of two main two steps. Each gene (DNA) must be converted, or transcribed, into messenger RNA (mRNA), RNA that serves as a template for protein synthesis. The resulting mRNA then guides the synthesis of a protein through a process called translation. Interestingly, mRNAs in a cell do not contain sequences from the regions between genes, nor from the non-coding introns that are present within many genes. Therefore, isolating mRNA is key to finding expressed genes in the vast expanse of the human genome. The problem, however, is that mRNA is very unstable outside of a cell; therefore, scientists use special enzymes to convert it to complementary DNA (cDNA). cDNA is a much more stable compound and, importantly, because it was generated from a mRNA in which the introns have been removed, cDNA represents only expressed DNA sequence.

> cDNA is a form of DNA prepared in the laboratory using an enzyme called reverse transcriptase. cDNA production is the reverse of the usual process of transcription in cells because the procedure uses mRNA as a template rather than DNA. Unlike genomic DNA, cDNA contains only expressed DNA sequences, or exons.

From cDNAs to ESTs

> A **"gene family"** is a group of closely related genes that produces similar protein products.

> A UTR is that part of a gene that is not translated into protein.

Once cDNA representing an expressed gene has been isolated, scientists can then sequence a few hundred nucleotides from either end of the molecule to create two different kinds of ESTs. Sequencing only the beginning portion of the cDNA produces what is called a 5' EST. A 5' EST is obtained from the portion of a transcript that usually codes for a protein. These regions tend to be conserved across species and do not change much within a gene family. Sequencing the ending portion of the cDNA molecule produces what is called a 3' EST. Because these ESTs are generated from the 3' end of a transcript, they are likely to fall within non-coding, or untranslated regions (UTRs), and therefore tend to exhibit less cross-species conservation than do coding sequences. A UTR is the part of a gene that is not translated into protein.

ESTs as Genome Landmarks

Just as a person driving a car may need a map to find a destination, scientists searching for genes also need genome maps to help them to navigate through the billions of nucleotides that make up the human genome. For a map to make navigational sense, it must include reliable landmarks or "markers". Currently, the most powerful mapping technique, and one that has been used to generate many genome maps, relies on Sequence Tagged Site (STS) mapping. An STS is a short DNA sequence that is easily recognizable and occurs only once in a genome (or chromosome). The 3' ESTs serve as a common source of STSs because of their likelihood of being unique to a particular species and provide the additional feature of pointing directly to an expressed gene.

ESTs as Gene Discovery Resources

> ESTs are powerful tools in the hunt for known genes because they greatly reduce the time required to locate a gene.

Because ESTs represent a copy of just the interesting part of a genome, that which is expressed, they have proven themselves again and again as powerful tools in the hunt for genes involved in hereditary diseases. ESTs also have a number of practical advantages in that their sequences can be generated rapidly and inexpensively, only one sequencing experiment is needed per each cDNA generated, and they do not have to be checked for sequencing errors because mistakes do not prevent identification of the gene from which the EST was derived.

> Using ESTs, scientists have rapidly isolated some of the genes involved in Alzheimer's disease and colon cancer.

To find a disease gene using this approach, scientists first use observable biological clues to identify ESTs that may correspond to disease gene candidates. Scientists then examine the DNA of disease patients for mutations in one or more of these candidate genes to confirm gene identity. Using this method, scientists have already isolated genes involved in Alzheimer's disease, colon cancer, and many other diseases. It is easy to see why ESTs will pave the way to new horizons in genetic research.

ESTs and NCBI

> For ESTs to be easily accessed and useful as gene discovery tools, they must be organized in a searchable database that also provides access to genome data.

Because of their utility, speed with which they may be generated, and the low cost associated with this technology, many individual scientists as well as large genome sequencing centers have been generating hundreds of thousands of ESTs for public use. Once an EST was generated, scientists were submitting their tags to GenBank, the NIH sequence database operated by NCBI. With the rapid submission of so many ESTs, it became difficult to identify a sequence that had already been deposited in the database. It was becoming increasingly apparent to NCBI investigators that if ESTs were to be easily accessed and useful as gene discovery tools, they needed to be organized in a searchable database that also provided access to other genome data. Therefore, in 1992, scientists at NCBI developed a new database designed to serve as a collection point for ESTs. Once an EST that was submitted to GenBank had been screened and annotated, it was then deposited in this new database, called dbEST.

dbEST: A Descriptive Catalog of ESTs

Scientists at NCBI annotate EST records with text information regarding DNA and mRNA homologies.

Scientists at NCBI created dbEST to organize, store, and provide access to the great mass of public EST data that has already accumulated and that continues to grow daily. Using dbEST, a scientist can access not only data on human ESTs but information on ESTs from over 300 other organisms as well. Whenever possible, NCBI scientists annotate the EST record with any known information. For example, if an EST matches a DNA sequence that codes for a known gene with a known function, that gene's name and function are placed on the EST record. Annotating EST records allows public scientists to use dbEST as an avenue for gene discovery. By using a database search tool, such as NCBI's BLAST, any interested party can conduct sequence similarity searches against dbEST.

UniGene

Because a gene can be expressed as mRNA many, many times, ESTs ultimately derived from this mRNA may be redundant. That is, there may be many identical, or similar, copies of the same EST. Such redundancy and overlap means that when someone searches dbEST for a particular EST, they may retrieve a long list of tags, many of which may represent the same gene. Searching through all of these identical ESTs can be very time consuming. To resolve the redundancy and overlap problem, NCBI investigators developed the UniGene database UniGene automatically partitions GenBank sequences into a non-redundant set of gene-oriented clusters. Although it is widely recognized that the generation of ESTs constitutes an efficient strategy to identify genes, it is important to acknowledge that despite its advantages, there are several limitations associated with the EST approach. One is that it is very difficult to isolate mRNA from some tissues and cell types. This results in a paucity of data on certain genes that may only be found in these tissues or cell types. Second is that important gene regulatory sequences may be found within an intron. Because ESTs are small segments of cDNA, generated from a mRNA in which the introns have been removed, much valuable information may be lost by focusing only on cDNA sequencing. Despite these limitations, ESTs continue to be invaluable in characterizing the human genome, as well as the genomes of other organisms. They have enabled the mapping of many genes to chromosomal sites and have also assisted in the discovery of many new genes.

MOLECULAR MODELING

Proteins are fundamental components of all living cells. They exhibit an enormous amount of chemical and structural diversity, enabling them to carry out an extraordinarily diverse range of biological functions. Proteins help us digest our food, fight infections, control body chemistry, and in general, keep our bodies functioning smoothly. Scientists know that the critical feature of a protein is its ability to adopt the right shape for carrying out a particular function. But sometimes a protein twists into the wrong shape or has a missing part, preventing it from doing its job. Many diseases, such as Alzheimer's and "mad cow", are now known to result from proteins that have adopted an incorrect structure. Identifying a protein's shape, or structure, is key to understanding its biological function and its role in health and disease. Illuminating a protein's structure also paves the way for the development of new agents and devices to treat a disease. Yet solving the structure of a protein is no easy feat. It often takes scientists working in the laboratory months, sometimes years, to experimentally determine a single structure. Therefore, scientists have begun to turn toward computers to help predict the structure of a protein based on its sequence. The challenge lies in developing methods for accurately and reliably understanding this intricate relationship.

Levels of Protein Structure

Proteins function through their conformation.

To produce proteins, cellular structures called ribosomes join together long chains of subunits. A set of 20 different subunits, called amino acids, can be arranged in any order to form a polypeptide that can be thousands of amino acids long. These chains can then loop about each other, or fold, in a variety of ways, but only one of these ways allows a protein to function properly. The critical feature of a protein is its ability to fold into a conformation that creates structural features, such as surface grooves, ridges, and pockets, which allow it to fulfill its role in a cell. A protein's conformation is usually described in terms of levels of structure. Traditionally, proteins are looked upon as having four distinct levels of structure, with each level of structure dependent on the one below it. In some proteins, functional diversity may be further amplified by the addition of new chemical groups after synthesis is complete.

The stringing together of the amino acid chain to form a polypeptide is referred to as the primary structure. The secondary structure is generated by the folding of the primary sequence and refers to the path that the polypeptide backbone of the protein follows in space. Certain types of secondary structures are relatively common. Two well-described secondary structures are the alpha helix and the beta sheet. In the first case, certain types of bonding between groups located on the same polypeptide chain cause the backbone to twist into a helix, most often in a form known as the alpha helix. Beta sheets are formed when a polypeptide chain bonds with another chain that is running in the opposite direction. Beta sheets may also be formed between two sections of a single polypeptide chain that is arranged such that adjacent regions are in reverse orientation. The tertiary structure describes the organization in three dimensions of all of the atoms in the polypeptide. If a protein consists of only one polypeptide chain, this level then describes the complete structure. Multimeric proteins, or proteins that consist of more than one polypeptide chain, require a higher level of organization. The quaternary structure defines the conformation assumed by a multimeric protein. In this case, the individual polypeptide chains that make up a multimeric protein are often referred to as the protein subunits. The four levels of protein structure are hierarchal, that is, each level of the build process is dependent upon the one below it.

Proteins can be Dvided into To Gneral Casses Bsed on Teir Trtiary Sructure

- **Fibrous proteins** have elongated structures, with the polypeptide chains arranged in long strands. This class of proteins serves as major structural components of cells, and therefore their role tends to be static in providing a structural framework.
- **Globular proteins** have more compact, often irregular structures. This class of proteins includes most enzymes and most of the proteins involved in gene expression and regulation.

A protein's primary amino acid sequence is crucial in

determining its final structure. In some cases, amino acid sequence is the sole determinant, whereas in other cases, additional interactions may be required before a protein can attain its final conformation. For example, some proteins require the presence of a cofactor, or a second molecule that is part of the active protein, before it can attain its final conformation. Multimeric proteins often require one or more subunits to be present for another subunit to adopt the proper higher order structure. In any case, as we stated earlier, the entire process is cooperative, that is, the formation of one region of secondary structure determines the formation of the next region.

Allosteric Proteins

> **Allosteric proteins can change their shape and function depending on the environmental conditions in which they are found.**

Under certain conditions, a protein may have a stable alternate conformation, or shape, that enables it to carry out a different biological function. Proteins that exhibit this characteristic are called allosteric. The interaction of an allosteric protein with a specific cofactor, or with another protein, may influence the transition of the protein between shapes. In addition, any change in conformation brought about by an interaction at one site may lead to an alteration in the structure, and thus function, at another site. One should bear in mind, though, that this type of transition affects only the protein's shape, not the primary amino acid sequence. Allosteric proteins play an important role in both metabolic and genetic regulation. Traditionally, a protein's structure was determined using one of two techniques: X-ray crystallography or nuclear magnetic resonance (NMR) spectroscopy.

X-ray Crystallography

> When performing this technique, the molecule under study must first be crystallized, and the crystals must be singular and of perfect quality – a time-consuming and difficult task.

Crystals are a solid form of a substance in which the component molecules are present in an ordered array called a lattice. The basic building block of a crystal is called a unit cell. Each unit cell contains exactly one unique set of the crystal's components, the smallest possible set that is fully representative of the crystal. Crystals of a complex molecule, like a protein, produce a complex pattern of X-ray diffraction, or scattering of X-rays. When the crystal is placed in an X-ray beam, all of the unit cells present the same face to the beam; therefore, many molecules are in the same orientation with respect to the incoming X-rays. The X-ray beam enters the crystal and a number of smaller beams emerge: each one in a different direction, each one with a different intensity. If an X-ray detector, such as a piece of film, is placed on the opposite side of the crystal from the X-ray source, each diffracted ray, called a reflection, will produce a spot on the film. However, because only a few reflections can be detected with any one orientation of the crystal, an important component of any X-ray diffraction instrument is a device for accurately setting and changing the orientation of the crystal. The set of diffracted, emerging beams contains information about the underlying crystal structure. If we could use light instead of X-rays, we could set up a system of lenses to recombine the beams emerging from the crystal and thus bring into focus an enlarged image of the unit cell and the molecules therein. But the molecules do not diffract visible light, and X-rays, unlike light, cannot be focused with lenses. However, the scientific laws that lenses obey are well understood, and it is possible to calculate the molecular image with a computer. In effect, the computer mimics the action of a lens. The major drawback associated with this technique is that crystallization of the proteins is a difficult task. Crystals are formed by slowly precipitating proteins under conditions that maintain their native conformation or structure. These exact conditions can only be discovered by repeated trials that entail varying certain experimental conditions, one at a time. This is a very time consuming and tedious process. In some cases, the task of crystallizing a protein borders on the impossible.

Nuclear Magnetic Resonance (NMR) Spectroscopy

> Solution NMR is performed on a solution of macromolecules while the molecules tumble and vibrate with thermal motion.

The basic phenomenon of NMR spectroscopy was discovered in 1945. In this technique, a sample is immersed in a magnetic field and bombarded with radio waves. These radio waves encourage the nuclei of the molecule to resonate, or spin. As the positively charged nucleus spins, the moving charge creates what is called a magnetic moment. The thermal motion of the molecule—the movement of the molecule associated with the temperature of the material—further creates a torque, or twisting force, that makes the magnetic moment "wobble" like a child's top. When the radio waves hit the spinning nuclei, they tilt even more, sometimes flipping over. These resonating nuclei emit a unique signal that is then picked up on a special radio receiver and translated using a decoder. This decoder is called the Fourier Transform algorithm, a complex equation that translates the language of the nuclei into something a scientist can understand. By measuring the frequencies at which different nuclei flip, scientists can determine molecular structure, as well as many other interesting properties of the molecule. In the past 10 years, NMR has proven to be a powerful alternative to X-ray crystallography for the determination of molecular structure. NMR has the advantage over crystallographic techniques in that experiments are performed in solution as opposed to a crystal lattice. However, the principles that make NMR possible tend to make this technique very time consuming and limit the application to small- and medium-sized molecules.

Computational Modeling

Researchers have been working for decades to develop procedures for predicting protein structure that are not so time consuming and that are not hindered by size and solubility constraints. To do this, researchers have turned to computers for help in predicting protein structure from gene sequences, a concept called homology modeling. The complete genomes of various organisms, including humans, have now been decoded and allow researchers to approach this goal in a logical and organized fashion. Before we go any further, it is important to define some common terminology used in this field.

Common Terminology Used in Homology Modeling

- **Folding motifs** are independent folding units, or particular structures, that recur in many molecules.
- **Domains** are the building blocks of a protein and are considered elementary units of molecular function.
- **Families** are groups of proteins that demonstrate sequence homology or have similar sequences.
- **Superfamilies** consist of proteins that have similar folding motifs but do not exhibit sequence similarity.

Some Basic Theory

A computer-generated image of a protein's structure shows the relative locations of most, if not all, of the protein's thousands of atoms. The image also reveals the physical, chemical, and electrical properties of the protein and provides clues about its role in the body.

It is theorized that proteins that share a similar sequence generally share the same basic structure. Therefore, by experimentally determining the structure for one member of a protein family, called a target, researchers have a model on which to base the structure of other proteins within that family. Moving a step further, by selecting a target from each superfamily, researchers can study the universe of protein folds in a systematic fashion and outline a set of sequences associated with each folding motif. Many of these sequences may not demonstrate a resemblance to one another, but their identification and assignment to a particular fold is essential for predicting future protein structures using homology modeling. The scientific basis for these theories is that a strong conservation of protein three-dimensional shape across large evolutionary distances—both within single species, between species, and in spite of sequence variation—has been demonstrated again and again. Although most scientists choose high-priority structures as their targets, this theory provides the option to choose any one of the proteins within a family as the target, rather than trying to achieve experimental results using a protein that is particularly difficult to work with using crystallographic or NMR techniques.

Maximizing Results

Specific tasks must be carried out to maximize results when determining protein structure using homology modeling. First, protein sequences must be organized in terms of families, preferably in a searchable database, and a target must be selected. Protein families can be identified and organized by comparing protein sequences derived from completely sequenced genomes. Targets may be selected for families that do not exhibit apparent sequence homology to proteins with a known three-dimensional structure. Next, researchers must generate a purified protein for analysis of the chosen target and then experimentally determine the target's structure, either by X-ray crystallography and/or NMR. Target structures determined experimentally may then be further analyzed to evaluate their similarity to other known protein structures and to determine possible evolutionary relationships that are not identifiable from protein sequence alone. The target structure will also serve as a detailed model for determining the structure of other proteins within that family. In favorable cases, just knowing the structure of a particular protein may also provide considerable insight into its possible function.

Protein Data Bank

The PDB was the first "bioinformatics" database ever built and is designed to store complex three-dimensional data. The PDB was originally developed and housed at the Brookhaven National Laboratories but is now managed and maintained by the Research Collaboratory for Structural Bioinformatics (RCSB). The PDB is a collection of all publicly available three-dimensional structures of proteins, nucleic acids, carbohydrates, and a variety of other complexes experimentally determined by X-ray crystallography and NMR.

> PDB is supported by funds from the National Science Foundation, the Department of Energy, and two units of the National Institutes of Health: the National Institute of General Medical Sciences and the National Library of Medicine.

The Molecular Modeling Database

NCBI's Molecular Modeling DataBase (MMDB), an integral part of our Entrez information retrieval system, is a compilation of all of the PDB three-dimensional structures of biomolecules. The difference between the two databases is that the MMDB records reorganize and validate the information stored in the database in a way that enables cross-referencing between the chemistry and the three-dimensional structure of macromolecules. By integrating chemical, sequence, and structure information, MMDB is designed to serve as a resource for structure-based homology modeling and protein structure prediction

> MMDB records have value-added information compared to the original PDB entries, including explicit chemical graph information, uniformly derived secondary structure definitions, structure domain information, literature citation matching, and molecule-based assignment of taxonomy to each biologically derived protein or nucleic acid chain.

NCBI has also developed a three-dimensional structure viewer, called Cn3D, for easy interactive visualization of molecular structures from Entrez. Cn3D serves as a visualization tool for sequences and sequence alignments. What sets Cn3D apart from other software is its ability to correlate structure and sequence information. For example, using Cn3D, a scientist can quickly locate the residues in a crystal structure that correspond to known disease mutations or conserved active site residues from a family of sequence homologs, or sequences that share a common ancestor. Cn3D displays structure-structure alignments along with the corresponding structure-based sequence alignments to emphasize those regions within a group of related proteins that are most conserved in structure and sequence. Cn3D also features custom labeling options, high-quality graphics, and a variety of file exports that together make Cn3D a powerful tool for literature annotation.

PDBeast: Taxonomy in MMDB

Taxonomy is the scientific discipline that seeks to catalog

and reconstruct the evolutionary history of life on earth. NCBI's Structure Group, in collaboration with NCBI's taxonomists, has undertaken taxonomy annotation for the structure data stored in MMDB. A semi-automated approach has been implemented in which a human expert checks, corrects, and validates automatic taxonomic assignments. The PDBeast software tool was developed by NCBI for this purpose. It pulls text-descriptions of "Source Organisms" from either the original PDB-Entries or user-specified information and looks for matches in the NCBI Taxonomy database to record taxonomy assignments.

The Role of Taxonomy

- Taxonomy provides a vivid picture of the existing organic diversity of the earth.
- Taxonomy provides much of the information permitting a reconstruction of the phylogeny of life.
- Taxonomy reveals numerous, interesting evolutionary phenomena.
- Taxonomy supplies classifications that are of great explanatory value in most branches of biology.

COGs: Phylogenetic Classification of Proteins

The database of Clusters of Orthologous Groups of proteins (COGs) represents an attempt at the phylogenetic classification of proteins— a scheme that indicates the evolutionary relationships between organisms—from complete genomes. Each COG includes proteins that are thought to be orthologous. Orthologs are genes in different species derived from a common ancestor and carried on through evolution. COGs may be used to detect similarities and differences between species for identifying protein families and predicting new protein functions and to point to potential drug targets in disease-causing species. The database is accompanied by the COGnitor program, which assigns new proteins, typically from newly sequenced genomes, to pre-existing COGs. A Web page containing additional structural and functional information is now associated with each COG. These hyperlinked information pages include: systematic classification of the COG members under the different classification systems; indications of which COG members (if any) have been characterized genetically and biochemically; information on the domain architecture of the proteins constituting the COG and the three-dimensional structure of the domains if known or predictable; a succinct summary of the common structural and functional features of the COG members as well as peculiarities of individual members; and key references.

Detecting New Sequence Similarities: BLAST against MMDB

Comparison, whether of structural features or protein sequences, lies at the heart of biology. The introduction of BLAST, or The Basic Local Alignment Search Tool, in 1990 made it easier to rapidly scan huge databases for overt homologies, or sequence similarities, and to statistically evaluate the resulting matches. BLAST works by comparing a user's unknown sequence against the database of all known sequences to determine likely matches. Sequence similarities found by BLAST have been critical in several gene discoveries. Hundreds of major sequencing centers and research institutions around the country use this software to transmit a query sequence from their local computer to a BLAST server at the NCBI via the Internet. In a matter of seconds, the BLAST server compares the user's sequence with up to a million known sequences and determines the closest matches. Not all significant homologies are readily and easily detected, however. Some of the most interesting are subtle similarities that do not always rise to statistical significance during a standard BLAST search. Therefore, NCBI has extended the statistical methodology used in the original BLAST to address the problem of detecting weak, yet significant, sequence similarities. PSI-BLAST, or Position-Specific Iterated BLAST, searches sequence databases with a profile constructed using BLAST alignments, from which it then constructs what is called a position-specific score matrix. For protein analysis, the new Pattern Hit Initiated BLAST, or PHI-BLAST, serves to complement the profile-based searching that was previously introduced with PSI-BLAST. PHI-BLAST further incorporates hypotheses as to the biological function of a query sequence and restricts the analysis to a set of protein sequences that is already known to contain a specific pattern or motif.

> BLAST now comes in several varieties in addition to those described above. Specialized BLASTs are also available for human, microbial, and other genomes, as well as for vector contamination, immunoglobulins, and tentative human consensus sequences.

Structure Similarity Searching Using VAST

> The detection of structural similarity in the absence of obvious sequence similarity is a powerful tool to study remote homologies and protein evolution.

As just noted, a sequence-sequence similarity program provides an alignment of two sets of sequences. A structure-structure similarity program provides a three-dimensional structure superposition. Structure similarity search services are based on the premise that some measure can be computed between two structures to assess their similarities, much the same way a BLAST alignment is predicted. VAST or the Vector Alignment Search Tool, is a computer algorithm developed at NCBI for use in identifying similar three-dimensional protein structures. VAST is capable of detecting structural similarities between proteins stored in MMDB, even when no sequence similarity is detected. VAST Search is NCBI's structure-structure similarity search service that compares three-dimensional coordinates of newly determined protein structures to those in the MMDB or PDB databases. VAST Search creates a list of structure neighbors, or related structures, that a user can then browse interactively. VAST Search will retrieve almost all structures with an identical three-dimensional fold, although it may occasionally miss a few structures or report chance similarities.

The Conserved Domain Database

The Conserved Domain Database (CDD) is a collection of sequence alignments and profiles representing protein domains conserved in molecular evolution. It includes domains from SMART and Pfam, two popular Web-based tools for studying sequence domains, as well as domains contributed by NCBI researchers. CD Search, another NCBI search service, can be used to identify conserved domains in a protein query sequence. CD-Search uses RPS-BLAST to compare a query sequence against specific matrices that have been prepared from conserved domain alignments present in CDD. Alignments are also mapped to known three-dimensional structures and can be displayed using Cn3D (see above).

Conserved Domain Architecture Retrieval Tool

NCBI's Conserved Domain Architecture Retrieval Tool (CDART) displays the functional domains that make up a protein and lists other proteins with similar domain architectures. CDART determines the domain architecture of a query protein sequence by comparing it to the CDD, a database of conserved domain alignments, using RPS-BLAST. The Conserved Domain Architecture Retrieval Tool then compares the protein's domain architecture to that of other proteins in NCBI's non-redundant sequence database. Related sequences are identified as those proteins that share one or more similar domains. CDART displays these sequences using a graphical summary that depicts the types and locations of domains identified within each sequence. Links to the individual sequences as well as to further information on their domain architectures are also provided. Because protein domains may be considered elementary units of molecular function and proteins related by domain architecture may play similar roles in cellular processes, CDART serves as a useful tool in comparative sequence analysis.

> RPS-BLAST is a "reverse" version of PSI-BLAST, which is described above. Both RPS-BLAST and PSI-BLAST use similar methods to derive conserved features of a protein family. However, RPS-BLAST compares a query sequence against a database of profiles prepared from ready-made alignments, whereas PSI-BLAST builds alignments starting from a single protein sequence. The programs also differ in purpose: RPS-BLAST is used to identify conserved domains in a query sequence, whereas PSI-BLAST is used to identify other members of the protein family to which a query sequence belongs.

Application to Biomedicine

Although the information derived from modeling studies is primarily about molecular function, protein structure data also provide a wealth of information on mechanisms linked to the function and the evolutionary history of and relationships between macromolecules. NCBI's goals in adding structure data to its Web site are to make this information easily accessible to the biomedical community worldwide and to facilitate comparative analysis involving three-dimensional structure.

SINGLE NUCLEOTIDE POLYMORPHISM (SNPS): VARIATIONS ON A THEME

Wouldn't it be wonderful if you knew exactly what measures you could take to stave off, or even prevent, the onset of disease? Wouldn't it be a relief to know that you are not allergic to the drugs your doctor just prescribed? Wouldn't it be a comfort to know that the treatment regimen you are undergoing has a good chance of success because it was designed just for you? With the availability of millions of SNPs, biomedical researchers now believe that such exciting medical advances are not that far away. A SNP (pronounced "snip"), is a small genetic change, or variation, that can occur within a person's DNA sequence. The genetic code is specified by the four nucleotide "letters" A (adenine), C (cytosine), T (thymine), and G (guanine). SNP variation occurs when a single nucleotide, such as an A, replaces one of the other three nucleotide letters—C, G, or T.

> **Although many SNPs do not produce physical changes in people, scientists believe that other SNPs may predispose people to disease and even influence their response to drug regimens.**

An example of a SNP is the alteration of the DNA segment AAGGTTA to ATGGTTA, where the second "A" in the first snippet is replaced with a "T". On average, SNPs occur in the human population more than 1 percent of the time. Because only about 3 to 5 percent of a person's DNA sequence codes for the production of proteins, most SNPs are found outside of "coding sequences". SNPs found within a coding sequence are of particular interest to researchers because they are more likely to alter the biological function of a protein. Because of the recent advances in technology, coupled with the unique ability of these genetic variations to facilitate gene identification, there has been a recent flurry of SNP discovery and detection.

> **As a result of recent advances in SNPs research, diagnostics for many diseases may improve.**

Finding single nucleotide changes in the human genome

seems like a daunting prospect, but over the last 20 years, biomedical researchers have developed a number of techniques that make it possible to do just that. Each technique uses a different method to compare selected regions of a DNA sequence obtained from multiple individuals who share a common trait. In each test, the result shows a physical difference in the DNA samples only when a SNP is detected in one individual and not in the other. Many common diseases in humans are not caused by a genetic variation within a single gene but are influenced by complex interactions among multiple genes as well as environmental and lifestyle factors. Although both environmental and lifestyle factors add tremendously to the uncertainty of developing a disease, it is currently difficult to measure and evaluate their overall effect on a disease process. Therefore, we refer here mainly to a person's genetic predisposition, or the potential of an individual to develop a disease based on genes and hereditary factors. Genetic factors may also confer susceptibility or resistance to a disease and determine the severity or progression of disease. Because we do not yet know all of the factors involved in these intricate pathways, researchers have found it difficult to develop screening tests for most diseases and disorders. By studying stretches of DNA that have been found to harbor a SNP associated with a disease trait, researchers may begin to reveal relevant genes associated with a disease. Defining and understanding the role of genetic factors in disease will also allow researchers to better evaluate the role non-genetic factors—such as behavior, diet, lifestyle, and physical activity—have on disease. Because genetic factors also affect a person's response to drug therapy, DNA polymorphisms such as SNPs will be useful in helping researchers determine and understand why individuals differ in their abilities to absorb or clear certain drugs, as well as to determine why an individual may experience an adverse side effect to a particular drug. Therefore, the recent discovery of SNPs promises to revolutionize not only the process of disease detection but the practice of preventative and curative medicine.

SNPs and Disease Diagnosis

Each person's genetic material contains a unique SNP pattern that is made up of many different genetic variations. Researchers have found that most SNPs are not responsible for a disease state. Instead, they serve as biological markers for pinpointing a disease on the human genome map, because they are usually located near a gene found to be associated with a certain disease. Occasionally, a SNP may actually cause a disease and, therefore, can be used to search for and isolate the disease-causing gene. To create a genetic test that will screen for a disease in which the disease-causing gene has already been identified, scientists collect blood samples from a group of individuals affected by the disease and analyze their DNA for SNP patterns. Next, researchers compare these patterns to patterns obtained by analyzing the DNA from a group of individuals unaffected by the disease. This type of comparison, called an "association study", can detect differences between the SNP patterns of the two groups, thereby indicating which pattern is most likely associated with the disease-causing gene. Eventually, SNP profiles that are characteristic of a variety of diseases will be established. Then, it will only be a matter of time before physicians can screen individuals for susceptibility to a disease just by analyzing their DNA samples for specific SNP patterns.

SNPs and Drug Development

Using SNPs to study the genetics of drug response will help in the creation of "personalized" medicine.

As mentioned earlier, SNPs may also be associated

with the absorbance and clearance of therapeutic agents. Currently, there is no simple way to determine how a patient will respond to a particular medication. A treatment proven effective in one patient may be ineffective in others. Worse yet, some patients may experience an adverse immunologic reaction to a particular drug. Today, pharmaceutical companies are limited to developing agents to which the "average" patient will respond. As a result, many drugs that might benefit a small number of patients never make it to market. In the future, the most appropriate drug for an individual could be determined in advance of treatment by analyzing a patient's SNP profile. The ability to target a drug to those individuals most likely to benefit, referred to as "personalized medicine", would allow pharmaceutical companies to bring many more drugs to market and allow doctors to prescribe individualized therapies specific to a patient's needs.

SNPs and NCBI

> Most SNPs are not responsible for a disease state. Instead, they serve as biological markers for pinpointing a disease on the human genome map.

Because SNPs occur frequently throughout the genome and tend to be relatively stable genetically, they serve as excellent biological markers. Biological markers are segments of DNA with an identifiable physical location that can be easily tracked and used for constructing a chromosome map that shows the positions of known genes, or other markers, relative to each other. These maps allow researchers to study and pinpoint traits resulting from the interaction of more than one gene. NCBI plays a major role in facilitating the identification and cataloging of SNPs through its creation and maintenance of the public SNP database (dbSNP). This powerful genetic tool may be accessed by the biomedical community worldwide and is intended to stimulate many areas of biological research, including the identification of the genetic components of disease.

NCBI's "Discovery Space" Facilitating SNP Research

Records in dbSNP are cross-annotated within other internal information resources such as PubMed, genome project sequences, GenBank records, the Entrez Gene database, and the dbSTS database of sequence tagged sites. Users may query dbSNP directly or start a search in any part of the NCBI discovery space to construct a set of dbSNP records that satisfy their search conditions. Records are also integrated with external information resources through hypertext URLs that dbSNP users can follow to explore the detailed information that is beyond the scope of dbSNP curation. Reproduced with permission from Sherry ST, Ward MH, Kholodov M, Baker J, Phan L, Smigielski EM, Sirotkin K."dbSNP: the NCBI database of genetic variation." *Nucleic Acids Research*. 2001; 29:308-311. To facilitate research efforts, NCBI's dbSNP is included in the Entrez retrieval system which provides integrated access to a number of software tools and databases that can aid in SNP analysis. For example, each SNP record in the database links to additional resources within NCBI's "Discovery Space". Resources include: GenBank, NIH's sequence database; Entrez Gene, a focal point for genes and associated information; dbSTS, NCBI's resource containing sequence and mapping data on short genomic landmarks; human genome sequencing data; and PubMed, NCBI's literature search and retrieval system. SNP records also link to various external allied resources. Providing public access to a site for "one-stop SNP shopping" facilitates scientific research in a variety of fields, ranging from population genetics and evolutionary biology to large-scale disease and drug association studies. The long-term investment in such novel and exciting research promises not only to advance human biology but to revolutionize the practice of modern medicine.

Impact of Biotechnology and Its Societal Benefits

The effects of any new technology introduced on the scale anticipated for biotechnology extend beyond the factories and research centers influencing our everyday lives. Biotechnology has, for example, made it possible to detect, and in some cases treat, diseases such as sickle-cell anemia, Tay-Sachs disease, diabetes, and cystic fibrosis. Following initial concerns that genetic engineering could give rise to infectious organisms - the spread of which would be difficult to contain - a stringent set of guidelines was drawn up by the government and leading scientists in the mid-1970s to regulate research in this field. While it is not possible to eliminate completely the risk of a genetic engineering accident, the experience of the last ten or so years of research has indicated that the chances of constructing a disease-producing organism by accident are very remote. This is because such pathogens require an extremely complex set of distinct characteristics, and are effective only when all are present.

Containment of experiments is the key to safety. Microbiologists have gained valuable experience over many years in handling extremely dangerous natural organisms, such as smallpox virus and cholera bacteria. Physical containment (airtight chambers and sterilization of all equipment) is backed up by biological containment. The K-12 strain of E. coli used for the vast majority of experiments, although originating from the E. coli organism present in the human intestine, has become accustomed to laboratory "penthouse" conditions. These optimum conditions are provided by microbiologists concerned with minimizing variation in laboratory data. As a result, scientists have shown that such strains cannot survive in the harsher conditions of the human body or the external environment. Other approaches include the use of strains that specifically require for their survival chemicals not present in the human body. Thus, current research carried out under the strictest guidelines carries minimal risk to workers and the public at large. Concern has been voiced that biotechnology might increase the risk of biological warfare, and some have speculated that biologists today are stepping into the shoes of the nuclear physicists of 40 years ago. It is undoubtedly a daunting aspect of the deployment of biotechnology that will require continued vigilance. Improved genetic tests based on biotechnological advances can be used to track down criminals in assault cases based on the uniqueness of their DNA. Genetic counseling can provide advice on heritable diseases, and genetic screening of workers in possible risk industries is being considered. DNA probes are providing breakthroughs in early diagnosis of disease. As detection of genetic predispositions becomes more predictable, a great deal may be known at birth of an individual's prospects in life. The moral question then arises as to who has access to this information and how this will affect the individual's quality of life. Many people have voiced concern about biotechnology and genetic engineering. Scientists have considered the issue of safety over recent years. A special committee of the National Academy of Sciences specifically reviewed the issues on the introduction into the environment of organisms genetically engineered using *recombinant DNA* technology. They concluded that "there is no evidence that unique hazards exist either in the use of R-DNA technique or in the transfer of genes between unrelated organisms," and that "the risks associated with the introduction of R-DNA engineered organisms are the same kind as those associated with the introduction of unmodified organisms."

The committee concluded that rDNA techniques

constitute a powerful and safe new means for the modification of organisms for the benefit of animals and humans. They also stated that there is adequate scientific knowledge to guide the safe and prudent use of such organisms outside research laboratories. It will be essential that such issues are aired in public debate as the technology develops. Many countries are actively reviewing the safety and ethics of biotechnology research and its applications. Some countries have already established research guidelines for work on embryo transplantation, embryo research, and surrogate motherhood. Lawyers and the public at large will be required to face up to these and similar questions as the biosciences, and biotechnology in particular, move forward. Legal problems have already emerged regarding patent laws. In 1980, for example, a U.S. court overturned existing practice and ruled that genetically-engineered microbes may be patented. The potential benefits include solving world food shortages, and improvements in medicine, agriculture, and veterinary sciences.

We can confidently expect biotechnological solutions to many essential industrial processes that currently produce toxic effluents. An increasing role for biotechnology in environmental management will undoubtedly follow. Because the prospect of serious biohazards appears to be receding, it does not mean that strict regulation of the new technology should be relaxed. Provided such vigilance is maintained, mankind can look forward to a wide range of exciting prospects that stem from biotechnology.

Plant Genomics: Recent Advances

The practical impact of agricultural genomics in the new millennium is difficult to project. The sciences key to biotechnology and genomic technologies are developing quickly. Public and private sector roles and investments are clearly evolving. The capacity to identify and move genes across species barriers has moved ahead far faster than our understanding of gene function and the ways genetic modification may affect function at the molecular, cellular, and especially organismal and ecosystem levels. The policy arena is unsettled and the marketplace borders on chaotic. On a global basis GMOs have emerged as a universally pressing issue, linking North and South and farmers on all continents in a melting pot of science-trade-economic tensions. Contrary to the wishful thinking of some biotech advocates, public opposition in the United States to foods derived from GMOs is likely to grow as consumers learn more about the technology. The patent and intellectual property arena is a minefield and already is limiting the free exchange of germplasm, ideas, and research tools. Agricultural biotechnology and plant genomics clearly have great potential to help solve agricultural production and food security challenges. But people are asking, and will continue to ask, whether this promise is being realized. Exactly who stands to benefit and who might end up bearing the costs and unintended consequences as the agricultural biotechnology revolution forges ahead? Recent developments are not encouraging and explain in large part why public skepticism is growing. This skepticism may lead to new barriers and costs for those wishing to push the revolution onward. Today I share my sense of how the government and biotech industry are misreading public sentiment and concern. I suggest some practical steps that could help relieve growing tensions and broaden support for both public and private investments in agricultural genomics and biotechnology.

Role and Limits

One of the reasons many people worry about agricultural biotechnology is the assumption in some circles that just about any farm production or pest problem can be solved through genetic modification. Put in simplest terms, genes are destiny; get the genes right and any problem will disappear, or at least become much easier to manage. This notion rests on a dangerous lack of understanding of what goes on in an agricultural ecosystem and is one reason why biotechnology makes many people nervous. Farming entails the management of complex biological systems. Success depends upon managing crop and animal growth close to the limits of natural and biologic resources, given the heat units and moisture Mother Nature is kind enough to provide in a given season. Stress from all sources must be managed so that genetic yield potential is not lost, to the full extent that is economically practical. To reach this goal pests must be managed so that weed competition, insect damage and diseases are kept below economic threshold levels. Genetic improvement, whether through classical breeding or biotechnology, is no substitute for good judgement in the design of farming systems, nor for attentiveness and skill in day-to-day farm management.1 In a perfect world, resources available to bring about genetic improvement would be targeted principally toward removing those barriers to yield or crop quality that are inherently genetic or which can only be addressed through genetic modification because of economic considerations (e.g., overcoming rust diseases in small grain). Over most of the last four decades, this has been largely the case but since the early 1990s, there has been a steady shift of breeding emphasis toward compatibility with various pesticide- and toxin-based pest management technologies. For example, Illinois corn growers were offered 23 corn hybrids on the Pioneer website in early 1999. Of these, 10 were Bt-transgenic and three were herbicide tolerant . Pioneer listed 20 "value-added" attributes across the 23 corn varieties, of which 13 were pest management related (65 percent). Herbicide-tolerant soybean varieties were comparably prominent in Pioneer's 1999 soybean varietal offerings to Illinois farmers. One of my greatest concerns about agricultural biotechnology is that many companies are over-promising what genetic change can deliver and that farmers may come to believe that genetic improvement can take the place for stewardship and skill in managing farming systems. Asking too much of any genetic, chemical or biological technology is a near-universal recipe for trouble. Biotechnology does not change this reality. Ironically, the benefits delivered by a transgenic crop variety in a given farming system will often be inversely proportional to the scope of change brought about by it. A contemporary case in point — farmers are likely to lose a valuable and safe herbicide as a result of over-reliance on Roundup (glyphosate) Ready soybeans and corn.2 Glyphosate resistance in some weed species has already emerged and shifts in weed communities is undermining efficacy in some parts of the Cornbelt. Still, this technology could prove sustainable and profitable for both farmers and companies if relied on more modestly and managed in ways that avoid resistance and limit selection pressure on weed communities. If farmers lose Roundup as an effective herbicide, it will not be because of an inherent flaw in the molecule or even herbicide tolerant technologies. The failure will be tactical and traced to how the technology is incorporated within farming systems and managed across agricultural landscapes.

Applications in Food and Fiber Production

Some applications and aspects of biotechnology and genomics in agriculture are going to continue making people nervous:

- Applications inconsistent with proven principles of sound biologically based agricultural systems, for example herbicide tolerant plants and Bt-transgenics.

- Use of antibiotic marker genes and reliance on powerful viral promoters.
- Moving growth hormones from species to species.
- Technologies driven largely by an opportunity to market a new product in order to capitalize on intellectual property rights.

Studies continue to document already recognized problems with many GMO varieties like pest resistance, erratic performance, poor economic returns to farmers, and non-target and ecological impacts. Ongoing and future work is sure to uncover new and unforeseen problems since nature is never static. Biotechnology does not short-circuit the adaptive capacity of organisms and ecosystems and indeed in some instances likely will shift them into over-drive. Most consequences will be performance-related, but new food safety and health-related risks will emerge as well. Over time, scientists will piece together the genetic, physiological or ecological roots of such problems, a necessary step in building confidence in the regulatory process. Policy decisions in the next half-decade will determine whether and how this new information is used in distinguishing between beneficial technologies and those that are not worth the costs and/or risks they impose. Thus far, regulators have faced a manageable number of decisions. The number, scope, complexity and potential significance of applications submitted - and regulatory decisions to make - will grow exponentially in the years ahead. The U.S. government's policy infrastructure is already shaky, especially the so-called "coordinated framework" put in place in the mid-1980s. An increasing number of research articles on potential GMO risks are appearing in scientific literature. Those who continue to argue that there is no evidence or reason for caution will not be taken seriously. The same applies to biotech critics who paint with too broad a brush and dismiss all biotechnology as inherently evil. The media, markets, and society will begin to recognize and discount extreme messages. Still, a core challenge will remain - how can society form a reasonably stable consensus to move the science and technology forward, given how divisive these issues have become and how strongly many people feel, pro and con? Bringing a degree of balance to R+D objectives is one element of forming such a consensus. Right now in the agricultural arena, commercial interests and the conventional pesticide paradigm are clearly driving the technology. To date government has shown little will or capacity to foster a greater diversity in research approaches and paradigms. Some influential players and companies seem to appreciate that the time has come to begin hedging the enormous bet and investments already made on agriculture biotechnology. Others would "rather fight than switch." Eventually dialogue must replace confrontation. As this occurs, one useful exercise will be for promoters and skeptics to mutually identify applications of genomics and biotechnology in food production and processing that seem worthy of public and private investments despite whatever risks they may entail. Herein I discuss some candidates.

Biopesticide and Pest Management

Agriculture clearly needs new and safer tools to manage a range of insects and plant diseases. Farmers and pest management specialists must also recognize that narrow, specific solutions - whether cultural, chemical, or genetic - are universally vulnerable when relied upon unilaterally across much of an agricultural landscape. Progress is being made in the development and adoption of biointensive Integrated Pest Management systems (BioIPM), yet managing grower risk of major losses remains a major constraint. Fortunately, the pesticide industry has discovered and brought to market some very effective, much safer and selective biopesticides and closely related synthetic analogues. These "reduced risk" pesticides are key new tools for farmers and will help them move along the continuum toward increasingly "soft" and elegant BioIPM systems. One notable example is Spinosad, a new Dow AgroSciences bioinsecticide derived from the fermentation of an actinomycete species, Saccharopolyspora spinosa. It is not genetically engineered. Spinosad is already registered on over 100 crops and will soon be available to just about all fruit and vegetable growers. Last fall EPA granted a tolerance for spinosad on all "Raw Agricultural Commodities," an action that opens the door for Dow AgroSciences to obtain labels on just about every crop (see detailed information on tolerance setting on the Consumers Union "Food Quality Protection Act". Within a few years, millions of farmers worldwide will use it to reduce their reliance on risky, disruptive broad-spectrum insecticides. Managed carefully to avoid resistance and harm to bees, spinosad could become the most important and profitable insecticide ever discovered. The manufacturing and purification processes for spinosad are complex and the cost of the product per acre treated could constrain market growth. Dow AgroSciences is responding in several ways. The company has entered into a two-year research agreement with Biotica Technology, a U.K. company, which will use genomics and other biotech tools to find ways to improve the efficiency of the production process and yield of the active agent (Agrow, January 1, 2000). Progress will make it possible for the company to increase production and reduce per unit production costs, as well as costs per acre treated on the farm. This use of biotechnology to lower the cost of a promising biopesticide or reduced risk pesticide is hard to argue with. It points to the need for more systemic, focused effort in the discovery and manufacturer of low-risk and selective biochemical pesticides. Dow AgroSciences could further blur the line between biotechnology and "green" farming technologies by formulating a version of spinosad that meets the requirements of organic certification standards. This will require further innovation in formulation technology, since the list of organically-approved inert ingredients is limited. But the challenge is surely not insurmountable and would be welcomed in many quarters. The reduced risk fungicide azoxystrobin is another example of a highly promising and beneficial biopesticide. It is a synthetic analogue of naturally occurring fungal metabolites. Formerly Zeneca, now AstraZeneca, soon-to-be Syngenta brought azoxystrobin to market in the mid-1990s. The chemical has been a major market success, with global sales pushing $500 million annually. Other companies have related strobilurin products in their product development pipelines. Just as the case with spinosad, the only problems on the horizon are resistance management and bringing down the cost of strobilurin products to farmers. Genomics and biotechnology can and will help on both fronts. Work is underway in several companies to find more stable and active formulations that will extend the range of disease organisms controlled. Better understanding of the mechanisms giving rise to resistance development will be essèntial in devising effective resistance management strategies. Genomics will make it possible to unravel the genetic basis of resistance, perhaps soon enough to design and implement effective management strategies.

A potential role for jasmonic acid in triggering plant defenses is described in a fascinating 1999 paper in Nature. Most biopesticides work through some direct impact on the development, feeding behavior, energy metabolism or reproduction of the target pest. There are other, more elaborate mechanisms, however, such as triggering plants to emit chemicals that attract insect predators and parasitoids. Thaler was able to accomplish this goal in

California tomato fields. Applications of jasmonic acid triggered a response in the octadecanoid pathway, which appears to produce volatile chemicals that serve as a signal to certain insect predators. In this experiment caterpillar predators were attracted to treated plants and reduced feeding damage significantly. Several teams are also looking for ways to trigger or reinforce systemic acquired resistance (SAR), a widely studied mechanism through which a plant attains the ability to overcome pathogen infection and other pest threats.

There is already one chemical pesticide on the market designed to help trigger SAR (the Novartis product Actiguard, which contains the active ingredient acibenzolar-S-methyl). Oldroyd and Staskawicz have shown that transgene-induced SAR can broaden the spectrum of disease control possible through stimulation of SAR. Many companies are working on combined approaches that entail genetic modification to enhance SAR within a cultivar in conjunction with promoter genes linked in some way to the application of a chemical trigger, like jasmonic acid or the Novartis product Actiguard. While plant pathologists are focusing on ways to trigger or augment SAR, entomologists are pursuing a number of other strategies that reinforce or broaden the efficacy of biological control approaches.

A Japanese team increased the susceptibility of armyworms to baculovirus by engineering rice plants to express an entomopoxvirus gene encoding for a virus enhancing factor. The experiment was remarkably successful - the dose of baculovirus needed to infect 50 percent of armyworm larvae feeding on transgenic rice leaves was reduced 260-360 times compared to non-engineered plants. In general, it seems likely that genetic modifications that strive to amplify or otherwise reinforce the effectiveness of a natural plant defense mechanism are likely to prove more stable and less prone to pleiotropic impacts than transformations creating wholly new mechanisms. They should also be more acceptable to the public.

New pests and strains of disease organisms emerge annually in new environments. Some people believe more virulent pathogens, exotic species and hard-to-control insects are emerging as problems faster than ever before (Ho et al.,1998 2000). Some people hypothesize that biotechnology, intensive farming and global trade may be contributing to the increased pace of natural adaptation and the dispersal of pest species. Geminiviruses are an important contemporary example. Another symposium speaker, Dr. Roger Beachy, has worked on a team demonstrating the capacity of geminiviruses to exchange DNA; a recent Science article reported that over 1,000 (naturally) recombinant strains have been found, including some from species with very different DNA sequences. As the organism expands its host range and becomes more virulent, more and more farmers face losses. The article goes on to explain that the search is on for resistant cultivars and that classical and/or genetic engineering techniques may be used to create them. The reader is left with the impression that only genetic modification can solve the problem posed by geminiviruses, despite the fact that the organism has proven itself very capable of evolving around once-effective genetic solutions. Classical or engineered resistance is only one tactic available for use against geminiviruses. These viruses help from insects to reach crop fields, spread within them, and induce infection. For years farmers in Florida have controlled these and related viruses most successfully through a multi-tactic system that rests upon aggressive elimination of sources of innoculum and timely control of their insect vectors. Even if plant breeders develop new sources of resistance that can be moved into commercial cultivars, resistance will last only if combined with other tactics that will lessen the chance of infection and delay the emergence of new strains. Deployment of such tactics must remain at the core of farmer-driven resistance management, regardless of whether the cultivar comes from classical or transgenic breeding techniques. The tools of biotechnology have much to offer in the discovery of new tactics and their integration into BioIPM systems, yet these options continue to attract very modest levels of public and private sector R+D resources. This leaves our national effort to better manage pests like geminiviruses unbalanced and unnecessarily vulnerable. It also suggests that something other than the goal of finding affordable and stable solutions is driving the technology development process.

Ecological Impacts

By 1996 there were over 25 million acres in the U.S. planted in transgenic plant varieties. About this time limited independent research began in a few U.S. laboratories on possible ecological impacts. The scope of work supported by U.S. Department of Agriculture funding remains limited; the USDA has allocated less than $2 million annually in support of the development of ag-biotech risk assessment methods, clearly a modest sum compared to the billion-plus investment annually in agricultural biotechnology. The results of the first wave of GE-ecological studies in the United States (those carried out since 1996) are now appearing in scientific journals.3 Most rely heavily on genomic and biotechnology techniques. Many focus on the persistence of GMOs in various environments and their ability to exchange DNA with other organisms. These are among critical variables determining the likelihood and extent of horizontal gene transfer. According to Tappeser and colleagues at the Institute for Applied Ecology in Germany,

"Horizontal gene transfer is now recognized to be the main avenue of exchange of genetic material in the microbial world, and hence also of the exchange and spread of antibiotic resistance genes" (Tappeser et al., 1998;). Recent work has shown that the digestive systems of invertebrates and vertebrates - from worms to mice to man - is a likely place where such transfers can occur. The excellent 1998 review article by Tappeser et al. cites several studies documenting such transfers. In addition, it is also becoming clear that gene flows can occur in just a matter of minutes, even seconds (Brockmann et al., 1996) and that there are often multiple mechanisms that might come into play under variable conditions. Biolog GN metabolic fingerprinting was used by a team to assess the differences in the microbial communities colonizing otherwise similar conventional and transgenic alfalfa varieties. The results "suggest that transgenic plant genotype may affect rhizosphere microorganisms and that the methodology in this study may prove a useful approach for the comparison of bacterial communities." These techniques may prove useful for the analysis of the impacts of Bt-corn hybrids for rootworm control.

Biocontrol of Soil Pathogens

Microbial biocontrol of damaging soil microorganisms (hereafter called rhizobacteria) represents a vast and largely untapped frontier. It is too expensive for farmers to use fumigants and other toxicants to control soil pathogens in most major grain, row and field crops. Very few farmers in developing countries can afford such costly methods to deal with soil-borne pests. For this reason cultural practices, particularly crop rotations, have been crucial worldwide in managing common root pathogens. Success tends to occur when a combination of practices creates a soil environment sufficiently conducive to non-damaging or beneficial microorganisms, such that they collectively keep Pythium, Rhizoctonia, or take-all diseases from decimating root systems. In the Pacific Northwest, Dr. James Cook and colleagues have worked on microbial biocontrol of these common wheat root diseases for some 30 years. As the tools of biotechnology came along, progress accelerated. Still, many steps were necessary to progress from recognizing a potential microbial biocontrol mechanism to developing a means to take advantage of it. In August 1988, two members of the team published a paper showing the role of the antibiotic phenazine-1-carboxylate in suppressing take-all decline and other root pathogens. Genomic and GE techniques that would be regarded as primitive today were used to create mutant strains of bacteria expressing the genes producing the phenazine antibiotic. The importance of the natural Pseudomonas fluorescens strain 2-79 in the production of this antibiotic was documented in a 1990 paper. An August 1992 paper showed that the antibiotics emitted by naturally occurring rhizobacteria were critical to their ability to survive in Pacific Northwest wheat field soils and that furthermore, these bacteria can play a critical role in the suppression of root diseases. Next the team characterized a 6.5 kb fragment from Pseudomonas fluorescens, strain Q2-87, the region that accounts for the production of another phenazine (Phl) antibiotic of interest.

This key work showed that the 6.5 kb fragment contained the Phl biosynthetic locus and that insertions into this fragment affected Phl production in a variety of ways. Cooperative work with Swiss scientists further refined understanding of the gene fragment with the Phl locus and drew upon it to produce a probe to test soils for the presence of strains of bacteria producing Phl. Two phenotypically distinct groups were found in soils from the U.S., Europe and Africa, leading to the insight that the genes for production of Phl are highly conserved in nature. In 1997 the team published a method to quantify the production of Phl in vitro and in the rhizosphere. This methodology would make it possible for the team to screen a large number of indigenous strains of bacteria, seeking those with the characteristics needed to produce a competitive and effective microbial soil inoculant or biopesticide. Late in 1997, the team informed the USDA and EPA of its intent to carry out a small field scale test of a transgenic microbial pesticide.

The microbial pesticide was formed by cloning the Phl biosynthetic pathway from Pseudomonas fluorescens strain 2-79 into Pseudomonas fluorescens strain Q8R1-96, which was known to be a better root colonizer. On December 19, 1997 the EPA published a notice in the Federal Register announcing receipt of the notification and inviting public comments.

This application of biotechnology was the first supported by Consumers Union in official comments submitted to the EPA. The comments were submitted in January 1998. The only critical comment and suggestion involved the need to replace the antibiotic marker gene used in the original transformation. Consumers Union supported this application because it was-

- Largely compatible with proven Integrated Pest Management principles;
- Involved the exchange of genes between closely related microorganisms (indeed Dr. Cook and colleagues hypothesize that the same exchange of genetic material had likely occurred in the past through natural bacteria conjugation); and
- There was no evidence that the new strain would gain a major competitive advantage or create any secondary problems.

Other Applications

Work is also underway in an effort to develop transgenic plants for phytoremediation and molecular farming. The idea is to "take advantage of the ability of plant roots to absorb or secrete various substances." In a compelling article in the Proceedings of the National Academy of Sciences, Gleba and colleagues point out that up to 10 percent of the photosynthetically fixed carbon is secreted through roots and state that -

> "Intuition and limited published data suggest that root-secreted compounds have a wide spectrum of biological activities including protection against biotic and abiotic stresses...The unexplored hemical diversity of root exudates is an obvious place to search for novel biologically active compounds."

An August 1999 article in Science reviewed efforts to develop more salt tolerant crops through genetic engineering, an idea first proposed 20 years ago. A team led by Apse identified and overexpressed the Na+/H+ antiport gene in Arabidopsis, creating significantly more salt-tolerant plants. Other strategies will no doubt emerge to try to increase the production potential of high saline soils. The economics of such technologies will eventually have to be evaluated relative to a variety of proven methods to either reduce salt levels in soil or limit further deposition. Since soil salinity is highly correlated with irrigation water quality and management and cropping system design, it is likely that the most cost-effective and sustainable solutions will involve a heavy dose of prevention, in addition to perhaps some increases in crop salt tolerance. If scientists and breeders try to push salt tolerance too far they are likely to trigger pleiotropic effects that impose additional costs and complications on farmers. It is also worth noting the complex linkages across social goals in regions dealing with high saline soils. In most cases, such areas are arid and high quality drinking water is a valuable public good in limited supply. Reducing the salt content in freshwater resources is often a goal of water quality programs designed to improve human health and preserve fisheries and biodiversity. Sometimes high saline conditions are simply caused by grossly inefficient and excessive fertilization, a problem whose solution almost always pays for itself.

Impacts on Soil Quality

Soil quality is generally defined as the capacity of a soil to take in, store and purify water, to hold and recycle nutrients, to support a diverse and robust biotic community, and to suppress pathogens and other pests. A seminal 1998 study in Nature by van der Heijden and colleagues showed that the diversity of mycorrhizal fungi plays a key role in determining the productivity of soil ecosystems. An overview by Read in the same issue speculates that greater fungal biodiversity expands the range of mechanisms through which microbial interactions can help plants deal with various sources of stress and competition. A team in Denmark has shown that indigenous soil bacteria can serve as a sink for plasmid-borne antibiotic resistance traits from E. coli added to agricultural soils (Sorensen et al., 1999). Holmes and colleagues in Oregon studied the impacts on soil biota of the addition of genetically engineered bacteria used in ethanol production (Holmes et al., 1999). They conducted a microcosm study in sandy soil supporting wheat plants. An engineered strain of the bacteria Klebsiella planticola stimulated the numbers of bacterial and fungal feeding nematodes and resulted in wheat plant decline and death, whereas plants survived following addition of the non-engineered strain. Their experiment showed the importance of plant-bacteria-nematode interactions in determining the impacts of changes in soil microbial communities and that impacts need to be assessed across several trophic groups of organisms. Disease suppressive soils arise because of interaction between plants and soil organisms. Key work by Goodman, Handelsman and colleagues at the University of Wisconsin have documented phenotypes in several plants that seem better able to trigger the disease suppressive potential of certain soil microorganisms, particularly Bacillus cereus strains. Their work shows that it is not just the soil, nor plant genetics that give rise to disease suppression, but an elaborate series of communications that arise from organisms in the soil to the growing plant and back to soil organisms through root exudates. It is also clear that a range of climatic, nutrient, cropping, and soil structure issues play key roles in modulating the communication between soil organisms and plants. One implication is clear—breeders will be able to create cultivars more resistant to a given disease by moving into a commercial variety specific genes that amplify a given plant defense response. But field performance may prove spotty because there are typically many different soil pathogens competing for a chance to thrive, as well as a variety of conditions that can break down channels of communication between soil organisms and plants.

For this reason, the cost of dealing with root diseases through specific plant genetic modification may prove to be prohibitive. Instead, the best strategies may be those that diversify or reinforce below-ground microbial biocontrol mechanisms or those that generically strengthen plant defense systems. The first transgenic variety designed to exude a toxin through its root system - a Bt-corn developed by Monsanto — is scheduled to reach the U.S. market in crop year 2001 or 2002, pending regulatory approvals. Impacts of this technology on soil food- webs and plant health and productivity are the major new concerns raised by this Bt-corn, engineered to control corn rootworm during its larval stage in the soil. The soil quality and plant health consequences of transgenic plants will trigger extensive debate because of growing interest in the role of indigenous soil organisms in sustaining productive and profitable farming systems. Some people argue that it is not now possible to evaluate the soil quality impacts of transgenic cultivars because there is no baseline of knowledge from which to monitor potential adverse impacts. Indeed, some of the basic ways plants communicate through their root systems with soil microorganisms are just now being identified. Because of our lack of baseline knowledge, some experts argue against the widespread planting of such varieties. Others respond by stating that tillage, crop rotations, and pesticides, especially soil fumigants have been profoundly altering soil microbial communities since farming began, and that ignorance of soil quality impacts never previously stopped the technology adoption process.

This debate frames a host of critical science and policy questions. Consensus and solid answers will take years to crystallize because so little is known about soil microbial ecology and how farming systems impact soil health, in turn influencing plant health and productivity. In the world of policy and research priorities, we have still not come to grips with serious impacts, and in many cases, degradation of soil microbial communities over the last six decades. Thus many ask "Why hold biotechnology to a higher standard?" One reason would be a priori evidence that transgenic technologies might impact soil microbial communities in ways that are qualitatively different or irreversible relative to other technologies and farming practices, like tillage, fertilization, or irrigation. Genomics and related biotechnology research could produce critical information to determine whether a threshold level of qualitatively distinct impacts has been met. To do so new research must focus on -

- Characterizing the short-term and longer-run fluctuations in soil microbial and invertebrate communities as a result of root exudates and toxins in root tissue. This area of work needs to ultimately trace impacts on soil and terrestrial food-webs.
- Linking these fluctuations and food-web changes to nutrient cycling and bioavailability (especially of phosphorous), changes in pest complexes and pressure, the triggering of plant immune response, and impacts on non-target organisms.
- Monitoring the emergence of resistance in soil organisms, including non-target species, in order to detect unexpected shifts in species fitness and competition.

Regulatory Policy

Biotech proponents argue that genetic engineering differs little from conventional breeding and that it is more precise and poses no new or hidden risks. It is increasingly hard to reconcile these simplistic assertions with emerging scientific knowledge. Of course, some applications of biotechnology are very similar to classical breeding or bring about a combination of genetic traits that may have actually occurred in nature. But much of the work underway entails moving genes across species barriers. When the tools of biotechnology are used to move novel genes into an organism, that organism will always respond and adapt in a variety of ways. Short-term responses will largely affect whether the genetic alteration is stable and useful relative to the desired change and whether the transformation leads to the silencing or changes in he expression of other desirable genes. Longer-term impacts and cellular responses can lead to gene silencing, codon bias, functional instability, and a range of pleiotropic effects. Many of these adaptations will be benign or of no consequence in most and perhaps even all circumstances. But under conditions of drought stress, when a particular nutrient is over- or under-supplied, when it is hotter or cooler than normal, when certain pests attack plants, in the presence of certain bacteria, and especially, when there are combinations of the above, plant physiological and immune systems will be forced into a sort of hyper-drive. It is under such conditions that unexpected and possibly risky consequences of genetic transformation are most likely, leading to what might be called "stress induced pleiotropic impacts." Examples of stress induced pleiotropic effects have already been documented. Petunias engineered to produce salmon red flowers broke down under conditions of stress, producing progressively fewer salmon red flowers (Meyer et al., 1992). Roundup Ready cotton boll drop appears linked to weather-induced stress, as is uneven expression of Bt toxins in cotton plants. Heat-induced yield problems with Roundup ready soybeans also appear to be a stress induced pleiotropic effect (Coghlan, 1999).

Regulating GMO Foods

In the mid-1980s, the United States government adopted the "substantial equivalence" policy based on the notion that foods derived from GMOs should be regulated in terms of their characteristics, and any quantifiable differences from non-GMO foods, rather than according to how they were produced. For the last decade the FDA, USDA, and EPA have received and approved thousands of applications to field test GMOs. Few, if any, applications have been turned down because of concern over risk. Through their provocative 1999 commentary in Nature, Millstone and colleagues triggered a useful debate over the scientific basis and impacts of the "substantial equivalence" policy. This policy lies at the heart of U.S. and European regulatory review and approval processes. It is important to assess further the soundness of the substantial equivalence policy, an issue which arose prominently in the public hearings the FDA sponsored in the summer and fall of 1999. A key step in doing so is to determine, since the policy was put in place, whether science has produced new information and insights that suggests or demonstrates gaps in current regulatory testing and reviews. One way to do so is to review the GMO risk concerns taken into account by regulatory scientists and agencies in the U.S. in the mid- to late 1980s, when the policy was codified, in contrast to risk concerns addressed recently in the scientific literature. A comprehensive review would be a major undertaking; here, I focus just on recent scientific literature that sheds light on a subset of potential GMO risks. Several recent overview articles and overviews have summarized the major mechanisms through which a GMO crop or food might lead to unanticipated adverse effects, either in the environment or in humans.

The major concerns covered in Table 1 are not exhaustive. Several ecological risks are also of recent vintage and belong in a more complete analysis. The earliest identified citations and trends in citation over time are shown. All searches were done using High Wire on the Science Online website. Keywords used in the search are shown in the table. For the seven categories in Table 1, the data show that:

- Less than 10 percent of the total number of citations covering seven major areas of risk appeared before 1990. About seventy percent have appeared since 1994. Accordingly, the "substantial equivalence" policy was formed well before many of today's most important risk concerns had been studied to any significant degree.
- Ninety percent of the total citations on "gene

silencing" have appeared in the literature since 1994 and almost 60 percent in just the last two years. Clearly gene silencing qualifies as a ubiquitous phenomenon, and one which is attracting a lot of attention.

- Relatively few studies have explored some critical concerns, e.g. pleiotropic effects.
- The number of papers in some areas like codon bias and naked DNA is accelerating. If current trends continue knowledge in these areas should rapidly expand.

Control Issues and Implications

Public attitudes toward GMO foods are shaped by several factors. Some people worry about health risks to themselves, their families, or the public at large. Others worry about ecological impacts. Ethical and religious concerns lie behind the concerns of others. And last, many people are most worried about a set of economic and control issues - patents, intellectual property rights, globalization, questions about who is driving the R+D agenda, distribution of costs and benefits, and impacts on corporate structure and control. There seems little chance that science is going to alleviate most of the health and environmental risk concerns, at least not any time soon. Given that agriculture is inherently a disruptive and biologically dynamic process, there are bound to be applications of genetic engineering in the food and fiber sectors that lead to problems. Even when isolated and contained, such episodes will reinforce the notion that we do not know enough to predict or manage the consequences following widespread planting of a given transgenic crop or the raising of a transgenic animal. Likewise, there is no end in sight to worries over patents, intellectual property issues, and corporate consolidation. The reason is obvious -

people cringe at the thought that corporations might someday control their genetic destiny, or that of their children. They worry that companies will use this power to extract larger profits from medical treatments. As a handful of companies gain control over crop germplasm, similar concerns arise over corporate control over the food supply. In both cases there are already real-world circumstances and episodes reinforcing these concerns. There are bound to be many more in the next few years given the pressure on biotechnology companies to justify high stock market valuations and attract new investors. Even in the absence of new evidence of health and ecological risks, control issues will keep biotechnology in the news and in the public eye.

Patents and Genomics Information

The rush in the private sector to control gene technology makes people nervous. Last month the popular press in Great Britain covered a story that must have reinforced concerns over the patenting of human genes. The story involves an attempt by the U.S.-based biotechnology company Myriad Genetics to compel British doctors to use their more sensitive breast cancer screening technology. The Myriad Genetics system is based on the BRCA1 and BRCA2 genes, over which Myriad holds a patent. British health officials dispute the breadth of the patent and are concerned that the substantially higher cost of the Myriad screen will limit the number of woman who can afford the procedure ("U.S. Firm may Double Cost of U.K. Cancer Checks," James Meek, The Guardian, January 17, 2000). Without doubt, genomics will lead to many novel and more sensitive diagnostic and screening tools. Their applications and costs, however, will raise all sorts of practical and ethical questions and controversies. If the general public comes to believe that cashing in on intellectual property is inordinately driving how the technologies are used, or that the technologies are widening the gap between rich and poor, the public will force the system to change, one way or another. In the agricultural arena, companies are now pursuing patents that cover routine physiological pathways and plant defense system responses. Novartis has applied for two patents in Europe (WO 99/35910 and WO 99/35913) covering an observed beneficial yield response when Bt-transgenic corn is sprayed with a number of pesticides, including nicotinoids like imidacloprid and thiomethoxam, the neural feeding inhibitor pymetrozine, a carbamate insecticide, and various herbicides. The claims in the patent cover unanticipated synergistic interactions between applications of various pesticides and yields of Bt-corn and even herbicide tolerant plants. The implications of the information put forth in these patent applications are that-

- GMO crops may be more vulnerable to certain pests than non-GMO varieties;
- Transgenic varieties will, in some circumstances need additional pesticide treatments to achieve maximum yields, perhaps because of some pleiotropic effect on natural plant defense mechanisms in addition to impacts on non-target beneficial insects and microorganisms; and
- In the long run GMO varieties may become more reliant on a broader range of pesticides in order to deal with always-changing pest complexes.

It is conceivable that a company might gain a patent on a gene or set of genes that plays a key role in triggering or modulating systemic acquired resistance (SAR), the production of semiochemicals or antibiotics, senescence, or even something as basic as photosynthesis. The significance of such patents could be enormous, since companies might be able to extract royalty payments from all other companies and public sector laboratories working at the genetic level on the same pathway or immune system response. At a minimum, the constant consideration of intellectual property right issues will continue to impose costs on the system and slow down scientific progress and the exchange of germplasm and breeding and biotechnology tools. According to a 1999 survey of public sector plant breeders conducted by Steven Price of the University of Wisconsin, almost half the breeders reported they had experienced difficulty in obtaining genetic stocks from private companies; 45 percent said that new restrictions on the exchange of germplasm had interfered with their research; and, about one-quarter reported that such restrictions had gotten in the way of graduate training. According to Price, the need to negotiate "material transfer agreements" (MTAs) is commonly the problem: "Negotiating these [MTAs] will continue to be difficult until the public and private sectors agrees on a common culture...If every MTA starts with each side putting its most onerous terms forward, then the negotiations will continue to be slow...public sector breeders will be hurt in the near term...in the long run, companies will be hurt by a decrease in trained plant breeders...the public may be hurt by decreased genetic diversity resulting in fewer varietal choices."

Material Transfer Agreements are a problem in other fields. The Pasteur Institute considers mouse MTAs as "a big administrative burden," a view clearly shared by many other institutions and scientists. An effort is underway by the International Mammalian Genome Society to develop a simple, standardized mouse MTA that will lower costs and encourage free exchange of mouse research tools (Abbott, 2000). Progress dealing with the problem has been made in the United States — Dupont and the NIH have reached a mutually acceptable agreement over access to Dupont's patented Cre-lox technology that allows certain genes to be removed from specific cells. Problems over access to this technology persist in Europe and within the private sector. The race by Celera Genomics and Incyte to profit from human genome sequence data is widely covered in both the scientific and lay media (for example, see Butler and Smaglik, 2000). The companies are pursing different strategies: Celera wants to sell access to its proprietary genomics data in return for fees and future royalties; Incyte is relying on patents over key gene sequences to position itself in the path of technology development. Ultimately the success of such companies and approaches will depend on whether the public believes that the rewards they seek and win from the marketplace are commensurate with their contributions in advancing science and technology. It is hard to imagine people accepting a system of patents and intellectual property protection that allows selected companies to extract large profits from genomic discoveries made possible by public and private sector investment over many years.

Control of Research Agendas

The notion that most agricultural production and nutrition problems can be solved by moving a few genes around has clearly been very seductive. It has led Wall Street to invest billions in biotechnology and underwritten the largely hostile takeover of the seed industry by the pesticide industry. It has fundamentally changed the focus and conduct of science in land grant universities, where molecular biology and the pursuit of genetic solutions now dominate departments that used to carry out a wider range of research, including significant applied field and systems-based research. This enormous shift in resources and the focus of agricultural science has occurred so quickly that there has not been adequate time for much reflection on the sustainability or value of the resulting technologies.

The loss of the benefits of research that has been abandoned in favor of molecular approaches has similarly been unexamined. In most quarters enthusiasm over the possible benefits of agricultural biotechnology has been infectious and unbounded. It has led many people to accept on faith that the technologies would generate enough "value added" returns to justify - and pay for - the costs of bringing the technologies to market. Science and experience are catching up with the myths generated by blind faith in agricultural biotechnology. The true value of biotechnology lies in its ability to provide farmers and those working with them the insights and tools needed to become better managers of the biological processes on which food production depends. Genetic improvements will surely remain a critical part of farming system innovation but over-reliance will lead to problems. The ecosystem and evolutionary forces behind resistance, weed shifts, the interplay of pests and beneficials, soil quality and plant health, and the emergence of new strains of disease organisms are complex, powerful and ubiquitous. Commercial biotechnology must come to terms with these forces. Much of the value of biotechnology will flow from its ability to help farmers understand and shape these forces rather than its ability to render them moot. Many changes will be required for this paradigm shift to take hold, including a rather significant change in public and private research priorities.

Private Sector Priorities

Private companies have invested most heavily in Bt-transgenic and herbicide tolerant varieties - so-called input traits. These applications of biotechnology have dominated private investment because they were technically possible and offered a new way for companies to extract greater profits from intellectual property rights - both those covering pesticide active ingredients and genes and genetic engineering techniques. The market success of Roundup Ready (RR) soybeans has been unprecedented in several ways. RR soybean varieties gained market share faster than any other new trait ever introduced.

The technology triggered a price war among herbicide manufacturers that has brought down herbicide prices to soybean farmers by about one-third across the board. RR soybeans have earned Monsanto hundreds of millions in "technology fees" linked to the purchase of seed - a new way to profit from intellectual property and seed industry assets. Most of the pesticide-seed industry is now investing heavily in bringing new herbicide tolerant varieties to market, hoping to capture a little of the magic that Monsanto has enjoyed with RR soybeans. Yet the success of RR soybeans is already leading to trouble. Resistance and shifts in weed communities are occurring that require farmers to make multiple applications of Roundup in fields where just one application worked for a year or two. Many farmers planting RR soybeans this year will apply two applications of Roundup along with at least one and often two other herbicides. Since RR soybean systems rest almost exclusively on herbicides for weed management, both reliance and average use will increase, contrary to claims that farmers planting these varieties use less herbicides. Experience with Bt-transgenic corn engineered to combat the European corn borer has been similar although not quite as dramatic. The technology has been adopted quickly and generated major income for Monsanto, Novartis, and AgrEvo. Notwithstanding consumer and trade jitters, the future of Bt-corn is suspect because of a range of inter-related ecological and biological problems that are coming into focus - resistance, impacts on non-target organisms and soil microbial communities and food webs, and value to farmers. Purdue University is the latest land grant university issuing a blunt report to corn growers on the economics of Bt-corn (Hyde et al., 2000). Because corn borer infestations are episodic, the extra cost of Bt-corn is not likely to pay a dividend in three out of every four years in Indiana. The researchers discuss a variety of factors such as planting date, level of infestation, yield potential and corn prices that farmers should take into account when making the decision whether to plant a Bt-corn hybrid. The message is clear - there are circumstances in Indiana where farmers are wise to pay the extra cost for Bt-hybrids but many more acres are planted to Bt-corn than can be justified by economic returns. Field experience with both herbicide tolerant and Bt-transgenic varieties led to an important insight - the value added by input traits is bounded by a variety of factors. Under conditions of high pest pressure or in regions where other technologies or solutions are not viable, transgenic varieties can offer farmers new tools that are worth their added cost. But when these varieties are marketed as near-universal "magic bullets," many farmers end up paying more and getting less in return. Plus, over-reliance can set the stage for resistance and other longer-term problems.

Public Sector Priorities

Biotechnology research now dominates work in most biological science departments at land grant universities. A 1996 Iowa State study showed that public sector plant breeding research fell 2.5 scientist-years per annum from 1990 to 1994 (Frey, 1996). Most faculty in entomology, plant pathology, weed science, agronomy and horticulture departments are either pursing biotechnologies or carrying out research on how to integrate them into contemporary farming systems. A small but growing number of scientists are working on developing methods to understand their environmental and ecological impacts. Few are left to study alternatives and system-based solutions. Funding shortfalls in public universities have made faculty and departments more dependent on carrying out work that the private sector is willing to support directly, or research that is likely to lead to patentable discoveries that might generate a royalty income stream. In either event, increased biotech-focused effort has come largely at the expense of traditional and applied field research on the interactions of farm system management, resources, pests and crop genetics. The shift is ongoing. A new Dean of the College of Agriculture at the University of Minnesota decided recently to expand agricultural genomics and biotechnology research and plans to do so in part by cutting funding that supports the Minnesota Institute for Sustainable Agriculture, the Center for Alternative Plant and Animal Products, and the Water Quality Center. This proposal has caught people's attention in St. Paul and a campaign is taking shape to block the shift in resources. The basic argument against the shift is that research priorities on campus are already far too heavily tilted toward molecular biology and "quick-fixes" of problems with high-input agriculture. The debate could grow spirited because of the large and well-organized community of farmers, scientists and activists working to promote sustainable agriculture in the State. The Dean's proposal might have the unintended but constructive consequence of bringing into the open a long-simmering debate on the mission of land grant universities in general and the University of Minnesota in particular.

Steps to Build Public Confidence

Many biotechnology promoters argue that there are fundamental differences between Europe and America relative to public attitudes on biotechnology. These conclusions are, in my judgement, based on a combination of sloppy and/or biased survey techniques and wishful thinking. Recent surveys, including some carried out by the food industry, show clearly that as Americans learn more about biotechnology and GMO foods, their skepticism grows.5 Given what emerging science and field experience are telling us about the benefits and potential risks arising from GMO food technologies, consumer and farmer skepticism may soon turn to anger and activism. It already has on Wall Street. Many biotech promoters are fighting back. Almost daily both the lay press and science journals contain new sweeping pronouncements and guarantees that "all's well" with the biotech revolution. In the last few months, triggered in part by events in Seattle during the WTO meeting, companies have financed new PR campaigns and are working to undermine individuals or groups that have raised questions about the safety or benefits of biotechnology. Politicians have entered the fray. An otherwise even-tempered and thoughtful member of the U.S. Senate has said recently that all evidence of GMO ecological or health risks is "hogwash." Few people find such responses persuasive.

Indeed, such statements tend to just harden public concern, reinforce simplistic notions regarding the extent of safety assessments in the U.S., and embolden activists. Still, much of the industry is stuck in the denial phase of grieving over the now irreconcilable breach between the vision and realities of agricultural biotechnology. More rocky ground lies ahead and the industry and GMO technologies will likely incur added costs and face new hurdles. The costs associated with insuring against and dealing with liability for problems in the field is an example of what could become a significant new cost-center for agricultural biotech pioneers. It is important to add that a full accounting of developmental, regulatory, product stewardship, labeling, marketing and liability associated with most agricultural biotechnologies is yet to be completed. But when things settle down, total costs are likely to prove sobering for much of the industry. The value added by most current GMO technologies is already stretched to the breaking point. People battling a disease may be willing to pay twice or three times more for an improved treatment or more effective drug, but very few farmers will, because first they won't be able to afford such a jump in cast costs and second, they will almost always have and pursue cheaper alternatives The only way to reduce such costs is to earn public confidence through independent research and thoughtful, thorough and transparent assessment of the benefits, costs and risks posed by given technologies. When risks are documented, companies and/or government agencies must take decisive and prudent action to avoid or limit problems, even in the absence of all the data risk assessors would like to have. In moving ahead with transgenic technology, risk assessors must factor into the equation their lack of understanding of how ecosystems and organisms might adapt.

Precautionary approaches that include careful monitoring of actual field impacts for the first few years after commercialization are critical. Assurances of safety based on an absence of evidence of human health problems and Food and Drug Administration "approval" of existing GMO foods are not convincing. I know this assertion is contentious, but before you dismiss this judgement as ill-informed, biased, or paranoid, read for yourself the views of several FDA scientists in the early 1990s as they developed the agency's "Statement of Policy" and other documents on GMO foods. Several such documents were obtained under the Freedom of Information Act and lie at the heart of the lawsuit against the FDA brought by the Alliance for Bio-Integrity. Over the next several months and perhaps years there will be a series of analyses and news reports triggered by this litigation. Many will contrast the technical concerns raised by FDA scientists in the 1991-1992 period with actual FDA decisions and actions. The more thoughtful pieces will also assess recent science to see what it has to say about the risk concerns identified by FDA scientists. Based on my review of dozens of key research reports published in recent years in several fields, many of the scientists' concerns have been borne out and indeed in many respects reinforced. Put another way, it is getting harder and harder to argue with a straight face that scientific evidence published in recent years resolves all the concerns raised by FDA scientists in the early 1990s. For these reasons, changes in FDA policy and procedure should be expected. The process of policy change could start the process of calming public concerns, especially if coupled with both increased public funding in essential areas of research and labeling of products now in the marketplace. If the policy reform process remains as contentious as it has been in recent years, it will progress slowly and satisfy no one.

Several steps in addition to FDA and USDA policy reforms and more public sector research will be needed to restore farmer and public confidence in agricultural biotechnology. Companies need to stop marketing GMO varieties as stand-alone solutions to complex problems with roots in farming system design and management. Marketing programs must emphasize when it is appropriate and inappropriate to select a transgenic variety, as well how the technology needs to be used to avoid resistance and adverse impacts on non-target organisms, limit the chances of gene flow, and assure solid economic returns to farmers. Both government and the private sector must more candidly and thoughtfully describe the limits of genetic technologies and gaps in understanding of potential risks. This need not stop a technology in its tracks but should lead to incremental introduction and trigger field monitoring and research on impacts. Scientists and regulators must be challenged to develop ways to distinguish almost assuredly safe and beneficial applications of transgenic technologies from those that raise significant risk concerns or offer dubious, at best short-lived benefits. There seems an ample enough supply of the former, promising applications, to justify delaying the commercialization of the risky ones. Acknowledging that there are legitimate risk-related questions is a first step. Funding independent researchers to pursue answers is a necessary second step. Scientists carrying out this work must be provided access to the tools and technologies needed to really understand how a plant or animal has been genetically transformed. They must not be punished personally or professionally if and when their results reinforce concerns, raise new questions, or point to the need for regulatory restrictions. Ways must be found to overcome the adverse impacts of patents and intellectual property policies on the conduct of science and the exchange, use, and improvement of germplasm. Fairness issues also must be dealt with.

The general public and world community has come to accept great disparity in wealth and income but the same cannot be said about access to and control over seeds and genetic resources. Pesticide companies with significant seed industry holdings bear a special burden to assure that their R+D and breeding priorities are not excessively weighted toward transgenic technologies linked to sale of their proprietary products in contrast to other solutions. Their actions and annual seed offerings will be carefully monitored in this regard. Agricultural biotechnology and genomics have gotten off to a rocky start. Its promise is now held hostage by problems the industry has largely brought upon itself. The credibility of both the U.S. government and industry is on the line, as is the public's soft spot for farmers and the economic performance of major sectors of the U.S. food system. Shifting gears in the current climate will require diplomacy and leadership. I applaud those in the biotechnology industry, agriculture, government, and the environmental and consumer communities who have started the process. Perhaps the single most constructive next step will be reaching agreement on at least some positive directions and applications and assuring that resources are invested in pursuing them.

3

Practical Aspects of Genetic Biotechnology

GENE GECHNOLOGY AND THE ENVIRONMENT: REGULATING RISKS AND BENEFITS

Gene technology involves altering the genetic 'blueprint' of living things to make them capable of producing new substances or performing new functions. Genes within a species can be modified, or these basic genetic building block of all organisms can also be transferred between species. Genetically modified organisms or *GMOs* have significant potential to benefit the environment. Using genetically modified plants and animals for agriculture, for example, could help reduce the environmental impact of farming. GMOs also have the potential for harmful impacts. Under the *Gene Technology Act 2000* a licence must be issued before a new GMO is released into the environment. Potential risks must be carefully identified, assessed, and managed to ensure that it does not pose a threat of serious or irreversible damage to the environment. The potential benefits and risks from gene technology, and the ways in which environmental risks are managed, are discussed below. The Department contributed funding to a large CSIRO research project to improve our understanding of the wider ecological impacts of GMOs.

The use of gene technology for managing the environment is at a very early stage, with many of the potential applications still being researched:

- The use of GMOs may provide more effective control of some exotic weeds and feral animal pests on land and in water. The CSIRO, for example, is developing immuno-contraceptive vaccines for foxes, rabbits and mice using gene technology.
- CSIRO and Orica Australia Ltd are using gene technology to develop enzyme products that detoxify pesticide residues, which in Australia would be of particular value to the cotton, horticultural and rice industries.
- Bacteria have been genetically modified as 'bioluminescors' that give off light in response to several chemical pollutants. These are currently being used to measure the presence of some hazardous chemicals in the environment.
- Other genetic sensors that can be used to detect various chemical contaminants are also being trialled. Some can be used to track how pollutants are naturally degrading in ground water.
- It may be possible in the future to include 'sterility' genes in plants and animals imported into Australia or specific regions so that they can breed only in captivity.
- The use of pest-resistant crops could mean a reduced use of insecticides, with less impact on other insects and animals, less pollution of air, waterways and soil, and reduced water and fuel use. The introduction of Bt cotton to Australia, which contains a toxin from the bacterium *Bacillus thuringiensis*, is already resulting in less use of insecticides to control bollworms.

- The use of genetically modified crops that are tolerant to broad-spectrum herbicides may result in more effective weed control, leading to less overall herbicide use and/or less demand for more narrowly targeted and persistent herbicides.
- Use of some genetically modified herbicide-tolerant crops may result in less cultivation for weed control thereby reducing the degradation of fragile soils.
- More efficient farming may lead to less pressure on the natural environment, including native bushland.

Before introducing new GMOs into the environment, regulators must assess a range of potential risks. Often there are strict conditions associated with the use of a GMO. If the environmental risks cannot be managed appropriately, then the organism is not approved for release. The following are examples of the types of risk considered.

- Immumocontraceptive GMOs could infect non-target native species.
- Overuse or misuse of herbicide-tolerant crops and pastures and the herbicides used on them could lead to herbicide resistant weeds, or sometimes referred to as 'super weeds'. Herbicide-tolerance genes could also transfer to wild populations or weedy relatives of the crop and increase the use of herbicides in non-agricultural areas.
- Insect-resistant crops could lead to insects developing new resistance. In the case of those crops where a bacterial gene is inserted to kill insect herbivores, resistant insects have already developed overseas in response to use of conventional, non-genetically modified products.
- Insect resistant crop plant material, for example from Bt cotton, may impact on soil ecosystems.

- In plants genetically modified to resist viruses, viral 'recombination' could create new viruses, with unknown environmental consequences.
- For genetically modified animals such as pigs or salmon, modifications for faster growth and larger size could transfer to native or feral populations should the GMO escape and interbreed. This could increase the fitness of feral animals, alter population dynamics or breeding patterns and disturb ecosystems.
- Tailoring crops to survive harsh conditions or for better pest or disease control may increase the scale and geographical extent of agriculture and its impact on the natural environment.

For more information see 'Submission by Environment Australia, Inquiry into Primary Producer Access to Gene Technology, House of Representatives Standing Committee on Primary Industries and Regional Services'.

The *Gene Technology Act 2000*, which came into force on 21 June 2001, regulates all dealings with genetically modified organisms in Australia. Under this legislation, licenses must be issued for any 'intentional release' into the environment. Before issuing such a licence, an independent Gene Technology Regulator must prepare a risk assessment and also a risk management plan. The Environment Minister is consulted both before and after these have been prepared. Written public submissions are also invited on the prepared risk assessment and risk management plan. In addition, the Department of the Environment and Heritage administers the *Environment Protection and Biodiversity Conservation Act 1999*. It regulates actions that are likely to have a significant impact on matters of national environmental significance, and also actions on Commonwealth land or actions by the Commonwealth that are likely to have a significant impact on the environment. The Act could be triggered by some dealings with genetically modified organisms.

Environmental Biotechnology, More Than Just Genetic Engineering

Biotechnology is the development of products or processes using plants, animals or micro-organisms. Sometimes this involves 'genetic engineering' (altering the genetic material of living things so that they can produce new substances or perform new functions). This fact sheet describes some of the many clever non-genetic engineering biotechnologies that can be used to protect the environment better and conserve biodiversity.

Cleaner Production

Environmental biotechnology offers ways to make industrial processes work more efficiently and create less pollution.

- Biotechnology may make available biological replacements for synthetic chemicals. For example, it is possible to make a variety of plastics from plant sugar rather than petrochemicals by using specially tailored yeasts and microorganisms. The advantage of these products is that they are biodegradable.
- It is expected that bioleaching will progressively displace chemicals used in recycling or bleaching paper. Many laundry detergents use enzymes (although in some cases these may be produced using gene technology) to replace phosphate detergents.
- Some heavy-duty stain removing bacteria have even

been found in heavily-alkaline lakes where they have survived by learning to break down the toxins in their environment.

- Scientists are looking for micro-organisms in hostile natural environments (for example very hot, cold or oily places) to use the talents they have developed for industrial uses.

Treatments for Solid Waste and Wastewater

- Most sewage treatment plants traditionally use a combination of chemical, physical and microbial treatment to break down waste. It is now possible to use modern biotechnology techniques to analyse the conditions needed to optimise the performance of the bugs and the systems, and to tailor the technology to different uses, from backyard septic tanks to large-scale intensive animal farm such as piggeries.
- The CSIRO has used enzymes to treat run-off from farms that contains pesticides, to avoid contaminating rivers and downstream farms.
- Sometimes the 'treatment' of waste produces valuable products, and modern high-tech composting factories can turn tonnes of organic garbage into precious soil in just days.

Bioremediation—Cleaning Up Contamination

- Naturally occurring micro-organisms, mainly bacteria and fungi, are being used to help clean up some of Australia's 60,000 sites contaminated by heavy metals, acids, petroleum derivatives, chlorinated solvents and explosives.
- Certain plants have also been found to absorb toxic metals such as mercury, lead and arsenic from

polluted soils and water, and scientists are hopeful that they can be used to treat industrial waste.

- Oil spills, from small industrial puddles to massive ocean oil spills are often treated with oil-eating bacteria.

Tracking the Health of the Environment Through Biomonitoring

Animals react to their environments, and can be used to gauge problems.

- The livers of sand flathead have proven to be useful indicators of municipal and industrial pollution from dioxins and other hazardous chemicals.
- Two frequently harvested species of marine mollusk, the Sydney rock oyster and the mussel, may be useful 'bioindicators' for the heavy metal toxins zinc and cadmium.

Biomass Energy

'Biomass' is plant and animal material that can be used as an energy source, from the traditional wood to waste material such as bagasse from sugarcane, to specially grown energy crops that can be converted to ethanol using modern biotechnology techniques and used with petrol in vehicles.

- Several Australian companies are planning to build biodiesel plants that convert vegetable oil or abattoir by-products into diesel fuel. Another company has a technology that converts green waste from rubbish tips into electricity.
- Even human waste can be converted into a combustible fuel.

Clever Plants

Genetic engineering is not the only tool for creating novel plants with special traits. Conventional breeding is still used for developing useful plants, and which can be done with fewer risks.

- By cross breeding, scientists are developing wheat plants to increase their salt tolerance so they can grow in areas of greater salinity.
- Eucalypts are also being cross-bred to introduce natural genes from other species to create plants that tolerate salinity and dryland conditions.

Genetic Engineering for Environmental Solutions

Many of the biotechnology applications described above can be varied or refined using genetic engineering.

- CSIRO and Orica Australia Ltd are using gene technology to develop enzyme products that detoxify pesticide residues, which in Australia would be of particular value to the cotton, horticultural and rice industries.
- In a variation on bioindicators, bacteria have been genetically modified as 'bioluminescors' that give off light in response to several chemical pollutants. These are currently being used to measure the presence of some hazardous chemicals in the environment.
- Other genetic sensors that can be used to detect various chemical contaminants are also being trialled. Some can be used to track how pollutants are naturally degrading in ground water.
- Many genetically modified plants have environmental benefits such as reduced pesticide use.

Gene Therapy

Each of us carries about half a dozen defective genes. We remain blissfully unaware of this fact unless we, or one of our close relatives, are amongst the many millions who suffer from a genetic disease. About one in ten people has, or will develop at some later stage, an inherited genetic disorder, and approximately 2,800 specific conditions are known to be caused by defects (mutations) in just one of the patient's genes. Some single gene disorders are quite common - cystic fibrosis is found in one out of every 2,500 babies born in the Western World - and in total, diseases that can be traced to single gene defects account for about 5% of all admissions to children's hospitals. In the U.S. and Europe, there are exciting new programs to 'map' the entire human genome - all of our genes. This work will enable scientists and doctors to understand the genes that control all diseases to which the human race is prone, and hopefully develop new therapies to treat and predict diseases.

Most of us do not suffer any harmful effects from our defective genes because we carry two copies of nearly all genes, one derived from our mother and the other from our father. The only exceptions to this rule are the genes found on the male sex chromosomes. Males have one X and one Y chromosome, the former from the mother and the latter from the father, so each cell has only one copy of the genes on these chromosomes. In the majority of cases, one normal gene is sufficient to avoid all the symptoms of disease. If the potentially harmful gene is recessive, then its normal counterpart will carry out all the tasks assigned to both. Only if we inherit from our parents two copies of the same recessive gene will a disease develop. On the other hand, if the gene is dominant, it alone can produce the disease, even if its counterpart is normal. Clearly only the children of a parent with the disease can be affected, and then on average only half the children will be affected. Huntington's chorea, a severe disease of the nervous system, which becomes apparent only in adulthood, is an example of a dominant genetic disease. Finally, there are the X chromosome-linked genetic diseases. As males have only one copy of the genes from this chromosome, there are no others available to fulfill the defective gene's function. Examples of such diseases are Duchenne muscular dystrophy and, perhaps most well known of all, hemophilia. Queen Victoria was a carrier of the defective gene responsible for hemophilia, and through her it was transmitted to the royal families of Russia, Spain, and Prussia. Minor cuts and bruises, which would do little harm to most people, can prove fatal to hemophiliacs, who lack the proteins (Factors VIII and IX) involved in the clotting of blood, which are coded for by the defective genes. Sadly, before these proteins were made available through genetic engineering, hemophiliacs were treated with proteins isolated from human blood. Some of this blood was contaminated with the AIDS virus, and has resulted in tragic consequences for many hemophiliacs. Use of genetically engineered proteins in therapeutic applications, rather than blood products, will avoid these problems in the future. Not all defective genes necessarily produce detrimental effects, since the environment in which the gene operates is also of importance. A classic example of a genetic disease having a beneficial effect on survival is illustrated by the relationship between sickle-cell anemia and malaria. Only individuals having two copies of the sickle-cell gene, which produces a defective blood protein, suffer from the disease. Those with one sickle-cell gene and one normal gene are unaffected and, more importantly, are able to resist infection by malarial parasites. The clear advantage, in this case, of having one defective gene explains why this gene is common in populations in those areas of the world where malaria is endemic.

Much attention has been focused on the so-called genetic metabolic diseases in which a defective gene causes an enzyme to be either absent or ineffective in catalyzing a particular metabolic reaction effectively. A potential approach to the treatment of genetic disorders in man is gene therapy. This is a technique whereby the absent or faulty gene is replaced by a working gene, so that the body can make the correct enzyme or protein and consequently eliminate the root cause of the disease. The most likely candidates for future gene therapy trials will be rare diseases such as Lesch-Nyhan syndrome, a distressing disease in which the patients are unable to manufacture a particular enzyme. This leads to a bizarre impulse for self-mutilation, including very severe biting of the lips and fingers.

The normal version of the defective gene in this disease has now been cloned. If gene therapy does become practicable, the biggest impact would be on the treatment of diseases where the normal gene needs to be introduced into only one organ. One such disease is phenylketonuria (PKU). PKU affects about one in 12,000 white children, and if not treated early can result in severe mental retardation. The disease is caused by a defect in a gene producing a liver enzyme. If detected early enough, the child can be placed on a special diet for their first few years, but this is very unpleasant and can lead to many problems within the family. The types of gene therapy described thus far all have one factor in common: that is, that the tissues being treated are somatic (somatic cells include all the cells of the body, excluding sperm cells and egg cells). In contrast to this is the replacement of defective genes in the germline cells (which contribute to the genetic heritage of the offspring). Gene therapy in germline cells has the potential to affect not only the individual being treated, but also his or her children as well. Germline therapy would change the genetic pool of the entire human species, and future generations would have to live with that change. In addition to these ethical problems, a number of technical difficulties would make it unlikely that germline therapy would be tried on humans in the near future. Before treatment for a genetic disease can begin, an accurate diagnosis of the genetic defect needs to be made. It is here that biotechnology is also likely to have a great impact in the near future.

Genetic engineering research has produced a powerful tool for pinpointing specific diseases rapidly and accurately. Short pieces of DNA called DNA probes can be designed to stick very specifically to certain other pieces of DNA. The technique relies upon the fact that complementary pieces of DNA stick together. DNA probes are more specific and have the potential to be more sensitive than conventional diagnostic methods, and it should be possible in the near future to distinguish between defective genes and their normal counterparts, an important development. The Human Genome Program in the U.S. will provide about $200 million each year to scientists in multidisciplinary research centers who are attempting to determine the makeup of all human genes. Together with similar programs in Europe, it is hoped that in 15 years time we shall be able to identify and treat all diseases to which humans are susceptible. This will revolutionize modern medicine, and hopefully improve the quality of life of all men, women, and children. Already, the genes for Duchenne muscular dystrophy, cystic fibrosis, and retinoblastoma have been identified, and more such information is emerging all the time.

Agricultural Biotechnology

Agricultural biotechnology includes, but is not limited to:

1. using our knowledge of cells to modify their activities.

2. modifying cells or microorganisms using the techniques of recombinant DNA rather than traditional selective breeding.
3. producing commercial quantities of desirable substances using microorganisms, cells or parts of cells.

Some examples of techniques used in agricultural biotechnology are:

1. selective breeding that produces crops and livestock with specified characteristics.
2. utilizing chemicals such as colchicine or radiation to induce mutations in genetic material.
3. growing new plants using specific parts of another plant in a process called vegetative propagation. Grafting is one example that is important in fruit trees.
4. encouraging the growth of a new plant in a prepared nutrient medium within a controlled environment.

Current examples of agricultural biotechnology products on the market are:

1. corn plants having resistance to a particular herbicide used to kill weeds.
2. corn plants producing a protein toxic to insect pests reducing the amount of insecticide use.
3. tomatoes producing less of a hormone that causes tomatoes to ripen thereby extending the shelf life.
4. corn hybrid plants growing in strongly alkaline soils that are found in the western US corn belt.
5. cotton plants resisting cotton bollworms and tobacco budworms.

A large percentage of watermelons imported into the US annually come from Mexico with the remaining volume arriving from Central and South America. Imports have increased to meet the consumer demand for watermelon all-year-round.

DNA Mediated Energy Transfer

The central component of all self-replicating molecular machinery alive today is the molecule known as Deoxyribonucleic acid (DNA). The molecular structure of DNA that allows it to translate, transcribe and copy information was first described by Watson and Crick in 1953 when they ended their *Nature* paper with the following understatement:

> "It has not escaped our notice that the specific pairing we have postulated immediately suggests a possible copying mechanism for the genetic material."

50 years of extrapolation on that idea have led to some of the greatest medical advances in history, leading eventually to the Human Genome Project. Considering that, imagine what the last 50 years may have brought had that sentence read:

> "It has not escaped our notice that the aromatic stacking that we have postulated immediately suggests a possible mechanism for nanoelectronic circuits."

The idea of charge transfers through DNA was indeed proposed in the years following 1953. However, since the electronic properties of DNA were not easily determined, it took about 40 years for the first experimental evidence of DNA mediated charge transfers to be published. Early skepicism claiming that the charge transfers occured though the backbone, or even through the surrounding solution have since been quenched by data showing that strong stacking within the double helix results in the fastest electron-transfer kinetics. Further research on electron transfers in DNA have indicated that the DNA sequence has predictable effects on the conductance of the ubiquitous biopolymer. Thus, depending on the sequence of the stretch of DNA propagating the electron transfer, DNA can either act as a *conductor*, a *semiconductor*, or an *insulator*. Most of the medically funded research on DNA mediated charge transfers focus on understanding how damaged DNA is repaired. For instance, when ionizing radiation or UV light strikes DNA, radicals can be formed within the DNA that may travel long distances before covalently binding atoms that are not normally connected. Studies involving both naturally occurring repair enzymes and mutagenically designed proteins are currently underway.

Nucleic Acids, Proteins and Lipid Membranes

Perhaps the most specific biomolecular self-assembly occurs when two single strands of Deoxyribonucleic Acid (DNA) assemble into the infamous double helix. Based on Watson and Crick base pairing, two complimentary strands of DNA will find each other within an enormously complex mixture. Double helical DNA strands can be separated from one another through simple heating, resulting in single stranded DNA (ssDNA). If the solution is allowed to cool, the two strands will re-assemble into a stable double helix. While the Watson and Crick base pairing then allows the specific assembly of the double helix, which is stabilized by relatively strong *hydrophobic* interactions between the aromatic bases within the center of the helix also known as *pi-stacking*. Nanoscientists can take advantage of the double helix's specific stability in numerous ways. One promising approach is to attach single-stranded DNA to small particles that one wishes to assemble. These components could range from diamondoid manifolds to metallic clusters that function as quantum dots. When DNA is used to create functional surfaces, the technique is called *DNA Directed Immobilization (DDI)*. Such patterned surfaces have applications for both nanoelectronic devices and biosensors. *Nanogen*, a biosensor manufacturer has a versatile method for that takes advantage of DNA's negative charge. By applying a voltage to various parts of a DNA containing solution, the concentrations of DNA over various parts surfaces can be regulated. As the primary components of the molecular machinery that makes life possible, proteins have a vast array of resources available to them for biomolecular self-assembly. Take for instance the eukaryotic nucleus of a multicellular organism (such as a human). The organization of the information contained within this tiny compartment is so vast that scientists are only now beginning to attempt to understand it. Containing the compartment is the nuclear membrane, which is filled with proteins that pump nutrients in and messages out. Entwined within what is called the "nuclear matrix" is the entire human genome, containing about 3 meters worth of DNA, packed into an area of only a few cubic micrometers. Keeping it all under control are then *histones* and thousands of other (non-histone) proteins that carry out processes such as *transcription* and *translation*. Each of these proteins has the appropriate information encoded within its peptide sequence for getting it to the location where it can do its job. In order for each of our cells to produce those proteins required for its specific task, only certain genes can be transcribed in certain cells. Thus, a very important amount of information is contained not only in the sequence of the DNA, but in its spatial organization within the nucleus. For instance, when the nucleus divides, all the contents must be unraveled, copied, and then reassembled back into two daughters of the original organized state. All this is accomplished by means of biomolecular self-assembly.

Structural Proteins *are generally called "fibrous" because they often have a fiber like nature. With names such as collagen, keratin, elastin, tubulin and fibroin, fibrous proteins self-assemble (often covalently) into large polymers that provide various nano- meso- and even macro-scale structures. Even our most modern synthetic polymers can not compete with fibrous proteins such as fibroin (silk) when it comes to strength, flexibility and overall molecular scale order. Thus, serious efforts are underway to better understand and take advantage of fibrous self-assembly.* Enzymes *(biological catalysts) are generally "globular" as opposed to "fibrous" and they often work together in the form of multi-enzyme complexes. This means that both intramolecular self-assembly and intermolecular self-assembly must take place in order for the enzyme to reach the conformation that allows it to do its job. Intramolecular self-assembly means that the newly synthesized protein chain must fold into a functional state. Most often this is accomplished by the aggregation of hydrophobic residues in the center of the protein, thus forcing hydrophilic residues to the surface of the protein. Once a single protein chain is adequately folded, it will often begin intermolecular self-assembly. In order to assemble a molecular machine known as a "multienzyme complex," the numerous subunits (or proteins) contact one another in ways that cause new surfaces to appear, thus allowing further subunit components to bind. Since all the information for this self-assembly is often contained within the protein sequence, it is possible to reproduce this assembly in a test tube. For instance, the molecular machine that builds proteins in E. Coli (the ribosome) can be taken apart into 58 isolated protein chains (as well as a few RNA molecules. If these components are mixed back together with the appropriate timing then functional ribosomes can be reassembled.* Membrane bound proteins *are incorporated into the cellular membranes that surround cells and their compartments. In order for these proteins to perform their numerous functions (signal transduction, ion balance maintenance, molecular transport, electron transport, structure, etc.), they must be appropriately oriented and organized within the membrane. Determining their structure within the membrane is not an easy task for biochemists, since the membranes are often destroyed during purification. However, membrane bound proteins often have a common structural characteristic that suggests a mechanism for their assembly. That is, several hydrophobic residues exposed on a certain area of the protein. Since the center of the membrane is hydrophobic, it can be expected that these hydrophobic residues stick to the inside of the membrane. Again, the hydrophobic effect is most effective for this purpose. Finally, once the molecule is incorporated into the membrane, changes in the membranes constitution as well as interactions with neighboring membrane bound proteins govern the organization of these proteins.* Cellular membranes define the boundries between various cellular compartments (such as the nucleus, mitochondria, endoplasmic reticulum, etc.) as well as enclosing the cell from the external environment. Similar to soap bubbles, cellular membranes are made of specific types of *surfactants* that tend to form bilayers. The primary surfactants are phospholipids, steroids (primarily cholesterol), and glycolipids. Two nanoscale bodies of water can be effectively separated from one by the formation of a bilayer of these molecules. The main barrier is the hydrophobic region in the center of a lipid bilayer membrane. Water and polar molecules dissolved in water can not pass through such a nonpolar barrier without the assistance of ion channels or molecular pumps.

Molecular Machines Protein Based

This section focuses on the study of the following main protein based molecular machines:

(i) ATP Synthase

(ii) The Kinesin, Myosin, Dynein and Flagella Molecular Motors

Synthesis of ATP is carried out by an enzyme, known as ATP Synthase. The inner mitochondrial membrane contains the ATP Synthase. The ATP Synthase is actually a combination of two motors functioning together as described. This enzyme consists of a proton-conducting F_0 unit and a catalytic F_1 unit. The figure also illustrates the subunits in side the two motor components. F_1 constitutes of $a_3b_3$gde subunits. F_0 has three different protein molecules, namely, subunit a, b and c. The g-subunit of F_1 is attached to the c subunit of F_0 and is hence rotated along with it. The a_3b_3 subunits are fixed to the b-subunit of F_0 and hence do not move. Further the b-subunit is held inside the membrane by a subunit of F_0 (shown in the figure by Walker).

1. Reversibility of the ATP Synthase: There are two directions in ATP Synthase system and these two directions correspond to two different functionalities and behavior. This two-way behavior is because of the reversible nature of the ATP-ADP cycle and the structure of the ATP Synthase. Let us term the forward direction as when the F_0 drives the g-subunit (because of proton motive force) of F_1 and hence ATP Synthesis takes place. And the backward direction is when hydrolysis of ATP counter-rotated the ã-subunit and hence the F_0 motor and leads to pumping back the protons. Therefore the forward direction is powered by the proton motive force and the backward direction is powered by the ATP hydrolysis. Which particular direction is being followed depends upon the situation and the environmental factors around the ATP Synthase.

2. Coupling of Proton Flow (F_0) and the ATP synthesis and hydrolysis (in F_1): Boyer proposed a model which predicted that the F_0 and F_1 motors are connected through the g-subunit. Further he proposed that this connection was mechanical in nature. The following figure illustrates the electrostatic surface potential on a_3b_3 subunits. The red colour represents a negatively charged surface and blue a positively charged surface.

3. Boyer's binding Change Mechanism: Boyer isolated the F_1 part of the ATP Synthase complex. It was found that the alpha and beta subunits alternate in this cylindrical part of the F_1 structure. As per this model each a and b pair forms a catalytic site. The rotation of the g subunit induces structural conformation in the a_3b_3 subunits. Although the three catalytic units are identical in their chemistry but they are functionally very different at any given point in time. These conformal changes induce a change in the binding affinities of all the three catalytic sites towards the ATPase reactants (ADP, P_i, ATP etc.).

The three catalytic sites could be in three distinct forms. O form stands for open which implies that it has got very low affinity for reactants; L form stands for loose, which implies that it would bind the reactants loosely but is still catalytically inactive; T form stands for tight, which binds the reactant tightly and is catalytically active. One of the sites being bound to ATP at T. The site for ADP and P_i would probably be L as it is more binding than the O site. Now when the energy is input to the system the sites are changed. T becomes O; O becomes L and L becomes T. Due to these changes the ATP which was previously in T is now released. Further the ADP and P_i are now tightly bound due to the conversion of L site to the T site. This tight binding of ADP and P_i allows them to spontaneously get converted into ATP. Hence, this model proposed that the proton flow from the F_0 is coupled to the site inter conversions in the F_1 unit which triggers synthesis-hydrolysis of ATP.

Boyer's theory was supported when Walker and his group solved the structure of the F_1-ATPase motor. The high resolution structure thus obtained gave hints towards many experiments which proved the fact that g subunit indeed rotated against the alpha and beta subunits.

4. F_1-ATPase a true nano rotary motor: Till today the exact mechanism of the molecular motor characterized by F_1-ATpase has not been fully determined. Research by Kinosita's lab is a step towards this goal and proposes some very conclusive models for the same. The results obtained show not only the various methods through which we can analyze these nano devices, but also predicts many characteristics for these.

What is known till now is that g subunit rotates inside the alpha-beta hexamer, but whether the rotation is continuous or is random was not known. Kinosita's lab solved this problem by imaging the F_1-ATPase molecule. The objective of their experiment was to determine the uniqueness of the rotary motion and its characteristics. The experiment that was performed by this group. They attached a micrometer long actin filament to the g subunit. This actin filament was fluorescently label, so that its fluorescence could be measured under a microscope. Hydrolysis of the ATP (when introduced in the experiment) led to the rotation of the g subunit and in effect the rotation of the actin filament. As reported by the authors, not all the actin filaments were observed to have rotation. But some percentage of them did rotate and that too in a unique direction and without having much reversibility in the direction. This direct imaging proved that the structure solved by Walker and group was indeed correct and there exists rotary motion between g subunit and the alpha and beta hexamer. The imaging of the actin filament that was obtained. The rotation is not restricted to some particular rotational angle but is continuously progressive is a particular direction. As the rotational motion is a continuous one, possibility of it being a twisting motion is ruled out. Hence, there should be no direct linkage between the g subunit and the alpha-beta hexamer in F_1-ATPase. It was further observed by the authors that the rotation of the actin filament was in certain steps. These step sizes were approximately equal to 120 degrees.

Rotational rates: Rotational rates were dependent upon the length of the actin filament. Rotational rates were found to be inversely proportional to the filament length. This could be attributed to the fact that the hydrodynamic friction is proportional to the cube of the length of the filament, therefore, longer length implied that the frictional forces were higher and rotational speeds in effect slower. As reported in the principle article, if the filament were to rotate at the speed of 6 rev per sec and the length of the actin filament were 1 mm, then as per the relation.

Efficiency of F_1: Work done in one step, i.e., 120 degree rotation is equal to the torque generated (=40 pN nm) times the angular displacement for that step (=2pi/3). This comes to about 80pN nm. Free energy obtained by the hydrolysis of one ATP molecule.

Here, the standard free energy change per ATP molecule = -50 pN nm at pH 7. Thermal energy at room temperature = 4.1 pN nm. At intracellular conditions, [ATP] and [P_i] is in the order of 10^{-3} M. Therefore, for [ADP] of 50 mM, DG = -90.604 pN nm. Comparing this value of the free energy obtained by the hydrolysis of one molecule of ATP with the mechanical work done by the motor (=80 pN nm), we can deduce that the F_1-ATPase motor works on efficiency close to 100 per cent!

- *Chemical synthesis of ATP powered by mechanical*

energy: In another study conducted by Hiroyasu Itoh, Akira Takahashi, Kengo Adachi, Hiroyuki Noji, Ryohei Yasuda, Masasuke Yoshida & Kazuhiko Kinosita Jr, evidence has been provided to justify the claim that the chemical synthesis of ATP occurs when propelled by mechanical energy. This basic nature of the F_1-ATPase was known for quite some while, but it is first time that it is being experimentally verified. To prove this concept, the F_1 was bound to the glass surface through histidine residues attached at the end of the g subunits. A magnetic bead coated with streptavidin was attached to the g subunit.

Electric magnets were used to rotate this bead attached to the g subunit. The rotation resulted in appearance of ATP in the medium (which was initially immersed in ADP). Thus, the connection between the syntheses of ATP as a result of the mechanical energy input is established. The exact mechanism of the F_1-ATPase rotation is still an active area of research today and many groups are working towards finding it. The key to solving the mechanism is solving the transient conformation of the catalytic sites and the g subunit when rotation is taking place. What is not clear is the correspondence between the chemical reactions at the catalytic sites and their influence on the rotation of the g subunit. Which event triggers the rotation and which not has still to be exactly determined? Many models have been predicted, but they all still elude the reality of the rotational mechanism. With modern microscopic tools, we view a cell as a set of many different moving components powered by molecular machines rather than a static environment. Molecular motors that move unidirectionally along protein polymers (actin or microtubules) drive the motions of muscles as well as much smaller intracellular cargoes. In addition to the F_0-F_1-ATPase motors inside the cell, there are linear transport motors present as tiny vehicles known as motor proteins that transport molecular cargoes that also require ATP for functioning. These minute cellular machines exist in three families—the kinesins, the myosins and the dyneins. The cargoes can be organelles, lipids or proteins etc. They play an important role in cell division and motility. There are over 250 kinesin-like proteins, and they are involved in processes as diverse as the movement of chromosomes and the dynamics of cell membranes. The only part they have in common is the catalytic portion known as the motor domain. They have significant differences in their location within cells, their structural organization, and the movement they generate. Muscle myosin, whose study dates back to 1864, has served as a model system for understanding motility for decades. Kinesin, however, was discovered rather recently using in vitro motility assays in 1985.

Conventional Kinesin is a highly processive motor that can take several hundred steps on a microtubule without detaching whereas muscle myosin executes a single "stroke" and then dissociates. A detailed analysis and modeling of these motors has been done. Kinesin and myosin make up for an interesting comparison. Kinesin is microtubule-based; it binds to and carries cargoes along microtubules whereas myosin is actin-based.

The motor domain of kinesin weighs one third the size of that of myosin and one tenth of that of dynein. Before the advent of modern microscopic and analytic techniques, it was believed that these two have little in common. However, the crystal structures available today indicate that they probably originated from a common ancestor. Myosin is a diverse superfamily of motor proteins. Myosin-based molecular machines transport cargoes along actin filaments—the two stranded helical polymers of the protein actin, about 5-9 nm in diameter. They do this by hydrolyzing ATP and utilizing the energy released. In addition to transport, they are also involved in the process of force generation during muscle contraction, wherein thin actin filaments and thick myosin filaments slide past each other. Not all members of the myosin superfamily have been characterized as of now. However, much is known about the structure and function. Myosin molecules were first sighted through electron microscope protruding out from thick filaments and interacting with the thin actin filaments in late 1950s. Since then it was known that ATP plays a role in myosin related muscle movement along actin. However, the exact mechanism was unknown, which was explained later in 1971 by Lymn and Taylor.

- *Structure of Myosin Molecular Motor:* A myosin molecule binding to an actin polymer. Myosin molecule has a size of about 520 kilodaltons (kD) including two 220 kD heavy chains and light chains of sizes between 15-22 kD. They can be visualized as two identical globular 'motor heads', also known as motor domains, each comprising of a catalytic domain (actin, nucleotide as well as light chain binding sites) and about 8 nm long lever arms. The heads, also sometimes referred to as S1 regions (subfragment 1) are shown in blue, while the lever arms or the light chains, in yellow. Both these heads are connected via a coiled coil made of two a-helical coils (grey) to the thick base filament. The light chains have considerable sequence similarity with the protein 'calmodulin' and troponin C, and are sometimes referred to as calmodulin-like chains. They act as links to the motor domains and do not play any role in their ATP binding activity but for some exceptions. The motor domain in itself is sufficient for moving actin filaments. Three-dimensional structures of myosin head revealed that it is a pear-shaped domain, about 19 nm long and 5 nm in maximum diameter.

- *Function of Myosin Molecular Motor:* A crossbridge-cycle model for the action of myosin on actin has been widely accepted since 1957. Since the atomic structures of actin monomer and myosin were resolved this model has been refined into a 'lever-arm model' which is now acceptable. Only one motor head is able to connect to the actin filament at a time, the other head remains passive. Initially the catalytic domain in the head has ADP and P_i bound to it and as a result, its binding with actin is weak. With the active motor head docking properly to the actin-binding site, the P_i has to be released. As soon as this happens, the lever arm swings counterclockwise due to a conformational change. This pushes the actin filament down by about 10 nm along its longitudinal axis. The active motor head now releases its bound ADP and another ATP molecule by way of Brownian motion quickly replaces it, making the binding of the head to the actin filament weak again. The myosin motor then dissociates from the actin filament, and a new cycle starts. However, nanomanipulation of single S1 molecules (motor domains) show that myosin can take multiple steps per ATP molecule hydrolyzed, moving in 5.3 nm steps and resulting in displacements of 11 to 30 nm.

Kinesin and Dynein family of proteins are involved in cellular cargo transport along microtubules as opposed to actin in the case of myosin. Microtubules are 25 nm diameter tubes made of protein tubulin and are present in the cells in an organized manner. Microtubules have polarity; one end being the plus (fast growing) end while the other end is the minus (slow growing) end. Kinesins move from minus end to plus end, while dyneins move from plus end to the minus end of the microtubules. Microtubule arrangement varies in different cell systems. In nerve axons, they are arranged longitudinally in such a manner that their plus ends point away from the cell body and into the axon. In epithelial cells, their plus end points towards the basement membrane. They deviate radially out of the cell center in fibroblasts and macrophages with the plus end protruding outwards. Like myosin, kinesin is also an ATP-driven motor. One unique characteristic of kinesin family of proteins is their processivity—they bind to microtubules and literally 'walk' on it for many enzymatic cycles before detaching. Also, each of the globular heads/motor domains of kinesin is made of one single polypeptide unlike myosin (heavy and light chains and dynein heavy, intermediate and light chains).

- *Structure of Kinesin Molecular Motor:* A lot of structural information about kinesin is now available through the crystal structures. The motor domain contains a folding motif similar to that of myosin and G proteins. The two heads or the motor domains of kinesin are linked via 'neck linkers' to a long coiled coil, which extends up to the cargo . They interact with the a and b-subunits of the tubulin hetrodimer along the microtubule protofilament. The heads have the nucleotide and the microtubule binding domains in them.
- *Function of Kinesin Molecular Motor:* While kinesin is also a two-headed linear motor, its modus operandi is different from myosin in the sense that both its head work together in a coordinated manner rather than one was being left out. The kinesin walk. Each of the motor heads is near the microtubule in the initial state with each motor head carrying an ADP

molecule. When one of the heads loosely binds to the microtubule, it looses its ADP molecule to facilitate a stronger binding. Another ATP molecule replaces the ADP which facilitates a conformational change such that the neck region of the bound head snaps forward and zips on to the head. In the process it pulls the other ADP carrying motor head forward by about 16 nm so that it can bind to the next microtubule-binding site. This results in the net movement of the cargo by about 8 nm. The second head now binds to the microtubule by losing its ADP, which is promptly replaced by another ATP molecule due to Brownian motion. The first head meanwhile hydrolyses the ATP and loses the resulting P_i. It is then snapped forward by the second head while it carries its ADP forward. Hence coordinated hydrolysis of ATP in the two motor heads is the key to the kinesin processivity. Kinesin is able to take about 100 steps before detaching from the microtubule while moving at 1000 nm/sec and exerting forces of the order of 5-6 pN at stall.

Dynein superfamily of proteins was introduced in 1965. Dyneins exist in two isoforms, the cytoplasmic and the axonemal. Cytoplasmic dyneins are involved in cargo movement, while axonemal dyneins are involved in producing bending motions of cilia and flagella.

- *Structure of Dynein Molecular Motor:* The structure consists of two heavy chains in the form of globular heads, three intermediate chains and four light intermediate chains. Recent studies have exposed a linker domain connecting the 'stem' region below the heads to the head itself. Also from the top of the heads the microtubule binding domains (the stalk region, not visible in the figure) protrude out. The ends of these stalks have smaller ATP sensitive globular domains which bind to the microtubules. Cytoplasmic dynein is associated with a protein complex known as dynactin, which contains ten subunits. Some of them are shown in the figure as p150, p135, actin related protein 1 (Arp1), actin, dynamitin, capping protein and p62 subunit. These play an important regulatory role in the binding ability of dynein to the microtubules. The heavy chains forming the two globular heads contain the ATPase and microtubule motor domains.

One striking difference that dynein exhibits compared to kinesins and myosins is that dynein has AAA (ATPases Associated with a variety of cellular Activities) modules, which indicate that its mode of working will be entirely different from kinesins and myosins. This puts dyneins into the AAA superfamily of mechanoenzymes. The dynein heavy chains contain six tandemly linked AAA modules with the head having a ring-like domain organization, typical of AAA superfamily. Four of these are nucleotide binding motifs, named P1-P4, but only P1 (AAA1) is able to hydrolyse ATP.

- *Function of Dynein Molecular Motor:* Since dynein is larger and more complex structure as compared to other motor proteins, its mode of operation is not as well known. However, very recently, Burgess et al., have used electron microscopy and image processing to show the structure of a flagellar dynein at the start and end of its power stroke; giving some insight into its possible mode of force generation. When the dynein contains bound ADP and V_i (vandate), it is in the pre-power stroke conformation. The state when it has lost the two, known as the apo-state is the more compact post power stroke state. There is a distinct conformational change involving the stem, linker, head and the stalk that produces about 15 nm of translation onto the microtubule bound to the stalk.

Unicellular organisms, such as, E. coli have an interesting mode of motility (for reviews see. They have a number of molecular motors, about 45 nm in diameter, that drive their 'feet' or the flagella that help the cell to swim. Motility is critical for cells, as they often have to travel from a less favorable to a more favorable environment. The flagella are helical filaments that extend out of the cell into the medium and perform a function analogous to what the oars perform to a boat. The flagella and the motor assembly are called a flagellum. The flagella motors impart a rotary motion into the flagella. In addition to a rotary mechanism, the flagella machines consist of components such as rate meters, particle counters, and gearboxes. These are necessary to help the cell decide which way to go, depending on the change of concentration of nutrients in the surroundings. The rotary motion imparted to the flagella needs to be modulated to ensure the cell is moving in the proper direction as well as all flagella of the given cell are providing a concerted effort towards it. When the motors rotate the flagella in a counterclockwise direction as viewed along the flagella filament from outside, the helical flagella create a wave away from the cell body. Adjacent flagella subsequently intertwine in a propulsive corkscrew manner and propel the bacteria. When the motors rotate clockwise, the flagella fly apart, causing the bacteria to tumble, or change its direction. These reversals occur randomly, giving the bacterium a 'random walk', unless of course, there is a preferential direction of motility due to reasons mentioned earlier. The flagella motors allow the bacteria to move at speeds of as much as 25 mm/s with directional reversals occurring approximately 1 per second. A number of bacterial species in addition to E. coli., depend on flagella motors for motility. Some of these are Salmonella enterica serovar Typhimurium (Salmonella), Streptococcus, Vibrio spp., Caulobacter, Leptospira, Aquaspirrilum serpens and Bacillus. The rotation of flagella motors is stimulated by a flow of ions through them which is a result of a build-up of a transmembrane ion gradient. There is no direct ATP-involvement; however the proton gradient needed for the functioning of flagella motors can be produced by ATPase.

- *Structure of the Flagella Motors:* A complete part list of the flagella motors may not be available as of now. Continued efforts dating back to early 1970s have however revealed much of their structure, composition, genetics and function. Newer models of the motor function are still being proposed with an aim to explain observed experimental phenomena. That means that we do not fully understand the functioning of this motor. A typical flagella motor from E. coli. consists of about 20 different proteins, while there are yet more that are involved in the assembly and functioning. There are 14 Flg-type proteins named FlgA to FlgN; 5 Flh-type proteins called FlhA to FlhE; 19 Fli-type proteins named FliA to FliT; MotA and MotB making a total of 40 related proteins. The name groups Flg, Flh, Fli and Flg originate from the names of the corresponding genes. Out of these the main structural proteins are FliC or the filament; FliD (filament cap); FliF or the MS-ring; FliG; FliM and FliN (C-ring); FlgB, FlgC and FlgF (proximal rod); FlgG (distal rod); FlgH (L-ring); FlgI (P-ring); FlgK and FlgL (hook-filament junction); and MotA-MotB (torque generating units). Earlier it was believed that the M and S are two separate rings and M was named after membrane and S after supramembranous. Now they are jointly called the MS-ring as it has been found that they are two domains of the same protein FliF. The C-ring is named after cytoplasmic, while the names of the P and L-rings come from 'peptidoglycan' and 'lipopolysaccharide' respectively. The FlhA,B, FliH,I,O,P,Q,R constitute the 'transport apparatus'.

The hook and filament part of the flagellum is located outside the cell body, while the motor portion is embedded in the cell membrane with parts (the C-ring and the transport apparatus) that are inside the inner membrane in the cytoplasmic region. MotA and MotB are arranged in a circular array embedded in the inner membrane, with the MS-ring at the center. Connected to the MS-ring is the proximal end of a shaft, to which the P-ring, which is embedded in the peptidoglycan layer, is attached. Moving further outwards, there is the L-ring embedded in the outer cell membrane followed by the distal shaft end that protrudes out of the cell. To this end there is an attachment of the hook and the filament, both of which are polymers of hook-protein and flagellin respectively.

- *Function of the Flagella Motors:* The flagellar motors in most cases are powered by protons flowing through the cell membrane (protonmotive force, defined earlier) barring exceptions such as certain marine bacteria, for example, the Vibrio spp., which are driven by Na+ ions. There are about 1200 protons required to rotate the motor by one rotation. A complete explanation of how this proton flow is able to generate torque is not available as of today. From what is known, the stator units of MotA and MotB play an important role in torque generation. They form a MotA/MotB complex which when oriented properly binds to the peptidoglycan and opens proton channels through which protons can flow. It is believed that there are eight such channels per motor. The protonmotive force is a result of the difference of pH in the outside and the inside of the cell. The E. coli cells like to maintain a pH of 7.6-7.8 on their inside, so depending on the pH of the surroundings, the protonmotive force will vary, and hence the speed of rotation of their motors. To test how the speed of rotation depends on the protonmotive force, the motors were powered by external voltage with markers acting as heavy loads attached to them. It was found that the rotation indeed depends directly on the protonmotive force. According to the most widely accepted model, MotA/MotB complex interacts with the rotor via binding sites. The passage of protons through a MotA/MotB complex (stator or torque generator) moves it so that they bind to the next available binding site on the rotor, thereby stretching their linkage. When the linkage recoils, the rotor assembly has to rotate by one step. Hence, whichever complex receives protons from the flux will rotate the rotor, generating torque. The torque-speed dependence of the motor has been studied in detail and indicates the torque range of about 2700 pN-nm to 4600 pN-nm.

As mentioned earlier, nature chose DNA mainly as an information carrier. There was no mechanical work assigned to it. Energy conversion, trafficking, sensing etc. were the tasks assigned mainly to proteins. Probably for this reason, DNA turns out to be a simpler structure—with only four kinds of nucleotide bases Adenosine, Thiamine, Guanine, and Cytosine (A, T, G & C) attached in a linear fashion that take a double helical conformation when paired with a complementary strand. Such structural simplicity vis-à-vis proteins—made of twenty odd amino acids with complex folding patterns—results in a simpler structure and predictable behavior. There are certain qualities that make DNA an attractive choice for the construction of artificial nano machines. In recent years, DNA has found use in not only mechanochemical, but nano electronic systems as well. A DNA double-helical molecule is about 2 nm in diameter and has 3.4-3.6 nm helical pitch no matter what its base composition is; a structural uniformity not achievable with protein structures if one changes their sequence. Furthermore, double stranded DNA (ds-DNA) has a respectable persistence length of about 50 nm which provides it enough rigidity to be a candidate component of molecular machinery. Single stranded DNA (ss-DNA) is very flexible and cannot be used where rigidity is required; however this flexibility allows its application in machine components like hinges or nanoactuators. Its persistence length is about 1 nm covering up to 3 base pairs at 1M salt concentration. Other than the above structural features there are two important and exclusive properties that make DNA suitable for molecular level constructions. These are—molecular recognition and self-assembly. The nucleotide bases A and T on two different ss-DNA have affinity towards each other, so do G and C. Effective and stable ds-DNA structures are only formed if the base orders of the individual strands are complementary. Hence, if two complementary single strands of DNA are in a solution, they will eventually recognize each other and hybridize or zip-up forming a ds-DNA. This property of molecular recognition and self-assembly has been exploited in a number of ways to build complex molecular structures. In the mechanical perspective, if the free energy released by hybridization of two complementary DNA strands is used to lift a hypothetical load, a force capacity of 15 pN can be achieved, comparable to that of other molecular machines like kinesin (5 pN). Dr. Nadrian Seeman and colleagues built the first artificial DNA based structure in the form of a cube in 1991. They then went on to create more complex structures such as knots and Borromean rings. In addition to these individual constructs, two dimensional arrays with the help of the double-crossover (DX) DNA molecule. This DX molecule gave the structural rigidity required to create a dynamic molecular device, the B-Z switch. DNA double helices can be of three types—the A, B or the Z-DNA. The B-DNA is the natural, right-handed helical form of DNA, while the A-DNA is a shrunken, low-humidity form of the B-DNA. Z-DNA can be obtained from certain C-G base repeat sequences occurring in B-DNA that can take a left-handed double helical form. The CG repeated base pair regions can be switched between the left and the right-handed conformations by changing ionic concentration.

The switch was designed in such a way that it had three cyclic strands of DNA, two of them wrapped around a central strand that had the CG-repeat region in the middle. On the two free ends of the side strands fluorescent dyes were attached in order to monitor the conformational change. With the change in ionic concentration the central CG-repeat sequence could alternate between the B and the Z modes bidirectionally which was observed through fluorescence resonance energy transfer (FRET) spectroscopy. In the year 2000, Dr. Bernard Yurke and colleagues made an artificial DNA based molecular machine that also accepted DNA as a fuel. The machine, called DNA tweezers, consisted of three strands of DNA labeled A, B and C. Strands B and C are partially hybridized on to the central strand A with overhangs on both ends. This conformation of the machine is the open conformation. When an auxiliary fuel strand F is introduced, that is designed to hybridize with both overhang regions, the machine attains a closed conformation. The fuel strand is then removed from the system by the introduction of its exact complement, leaving the system to go back to its original open conformation. This way, a reversible motion is produced, which can be observed by attaching fluorescent tags to the two ends of the strand A. In this case the 5' end was labeled with the dye TET (tetrachloro-fluorescein phosphoramidite) and the 3' end was labeled with TAMRA (carboxy-tertamethylrhodamine). Aside from the creation of a completely new molecular machine, this showed a way of selective fueling of such machines. The fuel strands are sequence-specific, so they will work on only those machines towards which they are directed, and not trigger other machines surrounding them. This machine was later improved to form a Three-State Device which had two robust states and one flexible intermediate state. A variation of the tweezers came about as the DNA-scissors.

In the past two decades, chemists have been able to create, modify and control many different types of molecular machines. Many of these machines carry a striking resemblance with our everyday macro-scale machines such as gears, propellers, shuttles etc. Not only this, all of these machines are easy to synthesize artificially and are generally more robust than the natural molecular machines. Most of these machines are organic compounds of Carbon, Nitrogen and Hydrogen, with the presence of a metal ion being required occasionally. Electrostatic interactions, covalent and hydrogen bonding play essential role in the performance of these machines. Such artificial chemical machines are controllable in various ways—chemically, electrochemically and photochemically (through irradiation by light). Some of them are even controllable by more than one ways, rendering more flexibility and enhancing their utility. A scientist can have more freedom with respect to the design of chemical molecular machines depending on the performance requirements and conditions. Rotaxanes and Catenanes make the basis of many of the molecular machines described in this section. These are families of interlocked organic molecular compounds with a distinctive shape and properties that guide their performance and control. Rotaxane family of molecular machines is characterized by two parts —a dumb-bell shaped compound with two heavy chemical groups at the ends and a light, cyclic component, called macrocycle, interlocked between the heads. It has been shown that a reversible switch can be made with a rotaxane setup. For this, one needs to have two chemically active recognition sites in the neck region of the dumb-bell. In this particular example, the thread was made of polyether, marked by recognition sites hydroquinol units and terminated at the ends by large triisoproplylsilyl groups. A tetracationic bead was designed and self-assembled into the system that interacts with the recognition sites.

The macrocycle has a natural low energy state on the first recognition site, but can be switched among the two sites reversibly upon application of suitable stimuli. Depending on the type of rotaxane setup the stimuli can be chemical, electrochemical or photochemical. The stereo-electronic properties of the recognition sites can be altered protonation or deprotonation, or by oxidation or reduction, thereby changing their affinity towards the macrocycle. In a recent example, ·light induced acceleration of rotaxane motion was achieved by photoisomerization, while similar controls through alternating current (oscillating electric fields) was shown before. The catenanes are also special type of interlocked structures that represent a growing family of molecular machines. They are synthesized by supramolecular assistance to molecular synthesis. The general structure of a catenane is that of two interlocked ring-like components that are non-covalently linked via a mechanical bond, that is, they are held together without any valence forces. Both the macrocyclic components have recognition sites that are atoms of groups that are redox-active or photochemically reactive. It is possible to have both rings with similar recognition sites. In such a scenario, one of the rings may rotate inside the other with the conformations stabilized by noncovalent interactions, but the two states of the inner ring differing by 180° will be undistinguishable (degenerate). For better control and distinguishable molecular conformations, it is desirable to have different recognition sites within the macrocycles. Then they can be controlled independently through their own specific stimuli.

The stereo-electronic property of one recognition site within a macrocycle can be varied such that at one point it has more affinity to the sites on the other ring. At this instant, the force balance will guide the rotating macrocycle for a stable conformation that requires that particular site to be inside the other macrocycle. Similarly, with other stimulus, this affinity can be turned off, or even reversed along with the affinity of the second recognition site on the rotating macrocycle increased towards those on the static one. There is a need for computational modeling, simulation and analysis of such molecular machine motion. Like rotaxanes, catenanes also can be designed for chemical, photochemical or electrochemical control. For both rotaxane and catenane based molecular machines, it is desirable to have recognition sites such that they can be easily controlled externally. Hence it is preferable to build sites that are either redox-active or photo-active. Catenanes can also be self assembled. An example of catenane assembled molecular motors is the electronically controllable bistable switch. An intuitive way of looking at catenanes is to think of them as molecular equivalents of ball and socket and universal joints. Pseudorotaxanes are structures that contain a ring-like element and a thread-like element that can be 'threaded' or 'dethreaded' onto the ring upon application of various stimuli. Again, the stimuli can be chemical, photochemical or electrochemical. These contain a promise of forming molecular machine components analogous to switches and nuts and bolts from the macroscopic world. Many other molecular devices have been reported in the past four decades that bear a striking resemblance to macroscopic machinery. Chemical compounds behaving as bevel gears and propellers that were reported in the late 1960s and early 1970s are still being studied today.

A molecular propeller can be formed when two bulky rings such as the aryl rings are connected to one central atom, often called the focal atom. Clockwise rotation of one such ring induces a counterclockwise rotation of the opposite ring about the bond connecting it to the central atom. It is possible to have a three-propeller system as well. Triptycyl and amide ring systems have been shown to observe a coordinated gear-like rotation. 'Molecular Turnstiles' which are rotating plates inside a macrocycle, have been created. Such rotations however were not controllable. A rotation of a molecular ring about a bond could be controlled by chemical stimuli, and this was shown by Kelly et al., when they demonstrated a molecular brake. A propeller-like rotation of a 9-triptycyl ring system, which was used in gears, this time connected to a 2, 2'-bipyridine unit could be controlled by the addition and subsequent removal of a metal. Thus, free rotations along single bonds can be stopped and released at will. Another type of molecular switch is the 'chiroptical molecular switch'. Another large cyclic compound was found to be switchable between its two stable isomeric forms P and M' (right and left handed) stimulated by light. Depending on the frequency of light bombarded on it the cis and the trans conformations of the compound 4-(2'-methoxythioxanthylidene)]-7-methyl-1,2,3,4-tetrahydrophenanthrene can be interconverted. Allowing a slight variation to this switch, a striking molecular motor driven by light and/or heat was introduced by Koumura and colleagues. As opposed to the rotation around a single bond in the ratchet described above, the rotation was achieved around a carbon-carbon double bond in a helical alkene. Ultraviolet light or the change in temperature could trigger a rotation involving four isomerization steps in the compound (3R,3'R)-(P,P)-trans-1,1',2,2',3,3',4,4'-octahydro-3,3'-dimethyl-4,4',-biphenanthrylidene. A second generation motor along with 8 other motors from the same material is now operational.

This redesigned motor has distinct upper and lower portions and it operates at a higher speed. This motor also provides a good example of how controlled motion at the molecular level can be used to produce a macroscopic change in a system that is visible to the naked eye. The light-driven motors when inside liquid crystal (LC) films can produce a color change by inducing a reorganization of mesogenic molecules. In this section we present two protein based motors that are at initial developmental stages and yet possess some very original and interesting characteristics. The idea of Viral Protein Linear motors stems from the fact that families of retroviruses like the influenza virus and the HIV-1 has a typical mechanism of infecting a human cell. When such a virus comes near the cell it is believed that due to the environment surrounding the cell it experiences a drop in pH of its surroundings. This is a kind of a signal to the virus that its future host is near. The drop of pH changes the energetics of the outer (envelope glycoprotein) protein of the viral membrane in such a way that there is a distinct conformational change in a part of it. A triple stranded coiled coil domain of the membrane protein changes conformation from a loose random structure to a distinctive a-helical conformation.

It is proposed to isolate this domain from the virus and trigger the conformational change by variation of pH in vitro. Once this is realized, attachments can be added to the N or C (or both) terminals of the peptide and a reversible linear motion can be achieved. In a recent development, large plant proteins that can change conformation when stimulated by positively charged ions were separated from their natural environment and shown to exert forces in orthogonal directions. Proteins from sieve elements of higher plants that are a part of the microfluidics system of the plant were chosen to build a new protein molecular machine element.

These elements change conformations in the presence of Ca2+ ions and organize themselves inside the tubes so as to stop the fluid flow in case there is a rupture downstream. This is a natural defense mechanism seen in such plants. The change in conformation is akin to a balloon inflating and extending in its lateral as well as longitudinal directions. These elements, designated as 'forisomes' adhered on to glass tubes were shown to reversibly swell in the presence of Ca^{2+} ions and shrink in their absence, hence performing a pulling or pushing action in both directions. Artificially prepared protein bodies like the forisomes could be a useful molecular machine component in a future molecular assembly, producing forces of the order of micronewtons. Unlike the ATP-dependant motors discussed previously, these machine elements are more robust because they can perform in the absence of their natural environment as well.

On Physics, Fundamentals, and Nanorobots: A Rebuttal to Smalley's Assertion that Self-Replicating Mechanical Nanorobots are Simply Not Possible

The September 2001 issue of *Scientific American* was devoted to nanotechnology. It had an article by Nobelist Richard E. Smalley in which he said: "Self-replicating, mechanical nanobots are simply not possible in our world." In support of this remarkable assertion, two principal objections are advanced: the "fat fingers" problem and the "sticky fingers" problem. "Both these problems are fundamental," Smalley writes, "and neither can be avoided." Here we show that neither problem represents a fundamental barrier that will prevent the construction of self-replicating, mechanical nanorobots. Further, Smalley's objections refer to designs that he himself proposed, not the designs of Drexler, whom he attempted to refute.

The "Fat Fingers" Problem

Smalley has previously stated that: "Chemistry is the concerted motion of at least 10 atoms." As this excluded a large part of well-known chemistry, it is not surprising that he now more circumspectly says: "In an ordinary chemical reaction five to 15 atoms near the reaction site engage in an intricate three-dimensional waltz that is carried out in a cramped region of space measuring no more than a nanometer on each side." The importance of the number of reactants lies in Smalley's argument that "There just isn't enough room in the nanometer-size reaction region to accommodate all the fingers of all the manipulators necessary to have complete control of the chemistry." Having trimmed the required number of fat fingers from ten to five, Smalley's argument is still on shaky ground. The argument collapses when we observe that chemical reactions often involve two reactants, such as in the controlled vacuum conditions used by the scanning tunneling microscope (STM). Two reactants can be brought together with controlled trajectories if one reactant is bound to a substrate and the second reactant is positioned and moved by a single "finger" as has already been done experimentally. For example, Ho and Lee physically bound a CO molecule to an iron atom on a silver substrate using an STM. Other approaches are also possible, Brenner *et al.* provided a molecular dynamics simulation of the hydrogen abstraction reaction from a diamond substrate. As noted elsewhere, if steric constraints near the tool tip make it unexpectedly difficult to manipulate particular individual atoms or small molecules with sufficient reliability, a simple alternative is to rely upon conventional solution or gas phase chemistry for the bulk synthesis of nanoparts consisting of 10-100 atoms. These much larger nanoparts can then be bound to a positional device and assembled into larger (molecularly precise) structures without further significant steric constraints. This is the approach taken by the ribosome in the synthesis of proteins. Individual amino acids are sequentially assembled into an atomically precise polypeptide without the need to manipulate individual atoms. "Atomically precise" is a description of the precision of the final product, not a description of the manufacturing method. Complete control of every aspect of a chemical reaction is not actually required to build a nanorobot. Effective control that delivers a precise product is what is necessary. The "fat fingers" problem is *not* a fundamental barrier to the development of molecular assemblers or the nanorobots they enable.

The "Sticky Fingers" Problem

Smalley also advances the "sticky fingers" problem, which is the claim that: "...the atoms of the manipulator hands will adhere to the atom that is being moved. So it will often be impossible to release this minuscule building block in precisely the right spot....these problems are fundamental...." The existence of some reactions that don't work still leaves plenty of room for reactions that do. To argue that the "sticky fingers" problem is a fundamental barrier to building mechanical assemblers and nanorobots, Smalley must show that no set of reactions exists which allows the synthesis of a useful range of precise molecular structures. To consider but one approach, application of a voltage between a manipulator tool and the workpiece can cause the target atom or moiety to move to the desired position. Again, Ho and Lee have provided an experimental existence proof. For a biological example, consider the ribosome.

This ubiquitous biological molecular assembler suffers from neither the "fat finger" nor the "sticky finger" problem. If, as Smalley argues, both problems are "fundamental," then why would they prevent the development of mechanical assemblers and not biological assemblers? If the class of molecular structures known as proteins can be synthesized using positional techniques, then why would we expect there to be no other classes of molecular structures that can be synthesized using positional techniques? Upon observing experimentally that polymers such as proteins can be synthesized under programmatic control, what convincing evidence do we have that the programmatic synthesis of stiffer polycyclic structures such as diamond is "fundamentally" impossible, and that mechanical assemblers will never be built? More directly, we can examine both those reactions that have been proposed specifically for use in a mechanical assembler and those reactions that take place using an SPM (Scanning Probe Microscope). *Nanosystems* discusses the mechanosynthetic reactions that might be used to synthesize some diamondoid structures of interest. Generally, these involve a single "finger," i.e., a probe tip with a functionalized end that would cause a site-specific reaction on a growing molecular work-piece. Merkle discusses several reactions which involve two, three, and even four reactants bound to the tips of molecular tools. A large and growing literature of research work with SPMs — both theoretical and experimental — supports the feasibility of site specific reactions involving a reactive tip structure interacting with a surface or with a molecule on a surface. In addition, SPMs are just the beginning of a growing list of tools for manipulating atoms and molecules.

Reductio ad Absurdum

One curious aspect of Smalley's comments is his stated replication time for a billion-atom "nanobot": one second. To obtain this blindingly fast replication speed, he adopts a 1 GHz atomic placement frequency — approximately one atom positioned every nanosecond. But Drexler's *Nanosystems*, the canonical textbook in this field, proposed a 1 MHz atomic placement frequency for mature molecular assembler systems with a corresponding replication time of ~10^3 seconds — figures that are commonly adopted in the supporting literature. Other proposals assume replication times up to ~10^5 seconds (~28 hours) for earlier and more primitive molecular assembler systems. Smalley's proposal of a one second replication time thus appears inconsistent with the extant literature, permitting him erroneously to extrapolate the replicative manufacture of a population of ~10^{18} nanorobots in just 60 seconds. This in turn leads to a rather obvious fallacy: Allowing ~500 zJ/bond and 1 bond/atom in Smalley's billion-atom molecular assembler, the power draw of a device population that occupies ~0.01 m^3 (~10^{27} atoms) during the last second of operation is ~1 GW, giving a power density of ~10^{11} W/m^3 which for diamond materials implies a mean autocatalytic graphitization decomposition time in vacuum of ~30 millisec, much briefer than the posited 1-second replication time — clearly a problem. In rhetoric, this approach might be called "*reductio ad absurdum*" or perhaps a "strawman." In a serious scientific discussion, a discrepancy of three orders of magnitude between what has been proposed in the literature and what is criticized suggests at best an inadequate grasp of the proposal. Smalley's article is accompanied by this categorical statement: "How soon will we see the nanometer-scale robots envisaged by K. Eric Drexler and other molecular nanotechnologists? The simple answer is never." As Arthur C. Clarke once observed: "When a distinguished but elderly scientist states that something is possible, he is almost certainly right. When he states that something is impossible, he is very probably wrong."

In spite of Clarke's observation, we recommend that rather than focus on issues of credentials, age, cabals, or orthodoxy, we simply stick to the scientific facts and their implications for molecular nanotechnology. The most direct way of evaluating the feasibility of engineering molecular assemblers and nanorobots is to conduct serious design and engineering efforts, along with detailed modeling and analysis. This would be the fastest way to reveal any hypothetical "fundamental problem" that would make non-biological molecular assemblers infeasible to build.

There are two possibilities: either molecular assemblers and the nanorobots they enable are theoretically feasible within the framework of well-understood existing physical laws, or they are not. A substantial and growing body of work strongly supports their feasibility. There are many worthy molecular systems engineering challenges to overcome, but thus far, there has been no credible argument that these devices are infeasible. Molecular nanotechnology assemblers would change society's relationship to molecules and matter as fundamentally as the computer changed our relationship to bits and information. If they are feasible, we need to focus research on development pathways and safeguards.

GAS SENSORS

MS Modeling's DMol^3 has been used to investigate the nitrogen dioxide sensing mechanism of tin dioxide nanoribbons. Understanding the sensing mechanism will enable the efficient design and manufacture of nanoscale chemical sensors—an important commercial application of nanotechnology. Practical challenges with carbon nanotubes involving cost of synthesis, control of chirality and diameter, separation from bundles, and attachment of functional groups have prompted researchers to explore other types of one-dimensional nanostructures. Two of the most promising candidates in this regard are nanowires and nanoribbons. As the name suggests, nanoribbons are solid objects (unlike nanotubes, which are hollow) with a near-uniform rectangular cross-section. So far, nanoribbons have primarily been synthesized from the oxides of metals and semiconductors. In particular, SnO_2 and ZnO nanoribbons have been materials systems of great current interest because of potential applications as catalysts, in optoelectronic devices, and as chemical sensors for pollutant gas species and biomolecules. Although they grow to tens of microns long, the nanoribbons are remarkably single-crystalline and essentially free of dislocations. Thus they provide an ideal model for the systematic study of electrical, thermal, optical, and transport processes in one-dimensional semiconducting nanostructures, and their response to various external process conditions. Recent experiments with SnO_2 nanoribbons indicate that these are highly effective in detecting even very small amounts of harmful gases like NO_2. Upon adsorption of these gases, the electrical conductance of the sample decreases by more than an order of magnitude. More interestingly, it is possible to get rid of the adsorbates by shining UV light, and the electrical conductance is completely restored to its original value. Such single-crystalline sensing elements have several advantages over conventional thin-film oxide sensors: low operating temperatures, no ill-defined coarse grain boundaries, and high active surface-to-volume ratio. To be able to fully commercialize their potential, it is important to better understand the sensing mechanism of such systems. With this goal in mind, researchers at the Brookhaven National Laboratory, Lawrence Berkeley National Laboratory, and Accelrys used MS Modeling's DMol[3] to investigate the nitrogen dioxide sensing mechanism of SnO_2 nanoribbons. Reporting in *Nano Letters*, the researchers examined the NO_2-sensing mechanism of SnO_2 nanoribbons with exposed (1 0'1) and (0 1 0) surfaces. The density functional theory (DFT) calculations revealed that:

- The most stable adsorbed species involved an unexpected NO_3 group doubly bonded to Sn centers
- An orders-of-magnitude drop in electrical conductance can be explained by significant electron transfer to the adatoms
- Computed binding energies were consistent with adsorbate stability up to 700 K, with X-ray absorption spectroscopy indicating predominantly NO_3 species on the nanoribbon surface.

The ability of the nanoribbons to sense O_2 and CO was also investigated. In the case of O_2, the response of the nanoribbon was highly sensitive to the concentration of O-vacancies on the surface. Thus, in the absence of any surface vacancies, the calculation predicted negligible charge transfer. However, when surface vacancies were present, an O_2 molecule can adsorb as a peroxide bridge, and withdraw a significant amount of electronic charge from the nanoribbon surface, thereby decreasing its electrical conductance. In the case of CO adsorption, there was a net electron transfer from the CO to the nanoribbon surface. Thus the calculation predicted an increase in nanoribbon electrical conductance upon CO-adsorption, in agreement with experimental results. Researchers from the Department of Energy's Sandia National Laboratories and the University of New Mexico (UNM) have developed a new way of mimicking photosynthetic proteins to manipulate platinum at the nanoscale.

The method has the potential of changing the metal's properties and benefiting emerging technologies. While we are in the early stages of research, we see the possibility of manipulating the nanoscale structure of platinum so that we can have control over the size, porosity, composition, surface species, solubility, stability and other functional properties of these metal nanostructures," says John Shelnutt, the Sandia scientist leading the research effort. "Such control means that the redesigned platinum could be used in many new applications, including catalysis, sensors and optoelectronic and magnetic devices." He adds that while research groups have reported a few platinum nanostructures — including nanoparticles, nanowires, nanosheets and others — the addition of new types of nanostructures is "highly desirable and potentially technologically important."

Working with Shelnutt in the research are Frank van Swol from Sandia, UNM graduate student Yujiang Song and Eulalia Pereira from the University of Porto in Portugal. The new method of manipulating platinum was detailed in a paper in the *Journal of American Chemical Society* published in December. The idea for the technique is similar to photosynthesis, in which plants use the energy from sunlight to produce sugar. But instead of manufacturing sugar, the new method changes a platinum ion to the neutral metal atoms. The photosynthetic protein mimicks this repeatedly, allowing metal to be deposited as desired at the nanoscale. The method involves putting porphyrins — the active part of photosynthetic proteins — along with the platinum salt in an aqueous solution of ascorbic acid at room temperature.

The porphyrins are placed in specific locations in the solution where it is intended that metal should be deposited. For example, the porphyrins may be confined to micelles, or liposomes. Micelles are spherical assemblies of detergent molecules in which the heads are exposed to the water and the tails stick together in the interior. Liposomes are similar structures, but they are larger and have water on the inside and outside separated by a closed membrane — sort of like a cell. The membrane is composed of two layers of detergent molecules, with the heads on the inner and outer surface facing the water and the tails forming the interior of the membrane. When light is shined on the porphyrins located in these detergent structures, the porphyrins excite, becoming catalysts for platinum reduction and deposition.

As this occurs, the metal grows onto the surfaces of the surfactant structures as a thin sheet or in other ways. In the case of micelles, the platinum grows into balls that look like the common toy "Koosh" ball. The ball size can be controlled by the amount of porphyrins and platinum in the solution, the amount of light illuminating the solution, and the amount of time the light is on.

For the metals platinum and palladium that form these nanostructures, it is enough for the porphyrin molecule to grow only a small metal "seed" particle composed of about 500 atoms. When it reaches this size, the seed starts to catalyze its own rapid growth (by oxidation of ascorbic acid), budding off arms in all directions and creating the Koosh-ball-like nanostructures. The porphyrin provides a convenient method of making these seeds at the location and time desired, leading to a uniform and selectable nanostructure size The platinum nanostructures take on a different form when they are prepared under different conditions. When the porphyrin is in a micelle, the platinum nanostructures produced look like Koosh balls. When the porphyrin is in the bilayer membrane of a liposome, the platinum grows in 2-nanometer thick sheet or platinum lace on the outer surface of the membrane, giving circular sheets — sort of like two-dimensional Koosh balls.

Under solution conditions for which the liposomes aggregate, growth can occur along the interfaces of the liposomes to give platinum foamlike materials and foam nanoballs. The type of nanostructure is mainly determined by the type of surfactant assembly upon which the platinum grows and the extent of growth from the individual seed nanoparticles. Since the porphyrin remains attached to the platinum nanostructure and active in the presence of light, it can also perform other functions besides growing itself.

For example, when illuminated with light, the platinum nanostructure evolves hydrogen from water. This reaction is similar to one of interest to car manufacturers looking for new ways to build automobiles powered by hydrogen fuel cells. Shelnutt says that in addition to structuring the platinum, the process also happens very fast. A few minutes in light will create many seeds, which then grow into the mature nanostructures in tens of minutes. And the process is easy to do.

Scientific Application Semiconductor Superlattices

A brief review of semiconductor superlattices is presented—their definition, basics of theory, types of materials and methods used for their preparation, the most important and unusual electrical and optical properties, which have already found their practical applications. When alternating layers of two semiconductors with different band gaps are deposited and thickness of these layers lies between 1-10 nm (which corresponds approximately to 10–100 atomic layers), what we obtain is semiconductor superlattice. The mostly known example of such a structure is GaAs/AlAs system depicted schematically together with its corresponding potential profile.

Differences in band gaps of materials lead to discontinuities in valence and conduction band and a set of square potential wells (low band gap semiconductor) separated by potential barriers (high band gap semiconductor) forms. Let us consider one of these wells from the quantum mechanical point of view. If we assume that an electron mean free path within a well is substantially longer than well width and the barrier width is infinite, what we obtain is a classical problem of a particle confined in a square well of finite depth. Unfortunately, in the real-life systems we cannot obtain perfect separation of adjacent wells and this results in coupling of their wave functions.

Thus, instead of discrete levels, this yields us subbands. An example of theoretical analysis of the GaAs/AlAs system shown, where an equal width, a, of well and barrier layer is assumed. Two main features common for other superlattice structures are easily seen when a is decreased. First, an energy distance between ground state and subsequent bound states increases, which happens because the well width is getting smaller. Second, the discrete levels split into the subbands as a result of thinner barrier layers and poorer separation of neighbouring wells. At this point, it is also important to notice, that the energy discontinuity, i.e. well depth, cannot be simply derived from such material constants as an energy gap, the conduction and valence band structure etc. in each semiconductor because of the lack of knowledge of the common energy reference. This parameter must be either calculated or measured. Historically, the crystalline superlattices where the first to produce, starting from pioneering work of Esaki and Tsu (1970). What is crucial in fabrication of such structures is nearly perfect match in their lattice constants. Otherwise the density of defects becomes so large that it results in strong electron scattering and obscures quantum phenomena. This is why a molecular beam epitaxy technique is utilised in their deposition and the number of materials, which can be combined to create superlattices, is restricted.

The best known and studied are GaAs/AlAs and GaAs/AsxGa1-xAs, InAs/GaSb are also worth mentioning. If interface defects cause such difficulties, why not to approach the problem from the opposite side—let us build the superlattice from the materials, where lattice constants no longer exist, i.e. amorphous semiconductors. This idea, apparently irrational, proved absolutely sensible in 1983, when Abeles and Tiedje reported in their paper fabrication of a-Si1-xNx:H/a-Si:H and a-Ge:H/a-Si:H systems. They stated that the strict requirements concerning lattice matching in crystalline superlattices can be relaxed due to amorphous structure by its very nature and a hydrogen content which plays an important role by passivating coordination defects.

Since then, many other materials were combined to produce superlattices : a-Si:H/a-C:H, a-Si:H/a-SiN:H, a-Si:H/a-SiNx, a-Si:H/a-SiOx, a-Si:H/a-Ge:H, CdSe/SeTe, CdSe/Se, a-Si3N4:H/a-Si:H, a-C:H superlattices where the band gap is tailored by changing deposition conditions. An impressive, comparing to the crystalline case, number of types of amorphous superlattices results from the fact that in this case we have much more freedom in choosing components of our systems as we are not limited by the lattice matching condition.It is also important to emphasise that the techniques utilised to produce amorphous films - plasma deposition, thermal evaporation, sputtering—are relatively cheaper than molecular beam epitaxy and easier to implement in industrial applications.

On the other hand, in case of amorphous multilayers it is more difficult to obtain abrupt interfaces between subsequent layers. Different methods are employed : pulsed application of microwave plasma during continuos r.f. glow discharge in Si2H6/N2, by which N2 is decomposed and incorporated into the growing film, nitridation of a-Si:H to attain a-Si3N4:H, lowering the growth rate in order to avoid fabrication of transition layers when the deposition conditions change, quick exchange of source gases. However, sometimes smooth transition from one layer to another is intentionally introduced, which may lead to sinusoidal potential profile, instead of a step-like. For the sake of convenience, electrical properties, particularly conduction phenomena, may be considered in terms of what is happening when we study currents flowing in-plane (parallel to the layers) or perpendicular to them.

As it has been stated above, quantum confinement causes shift of the first bound state allowed for charge carriers. This must of course result in the increase of an activation energy of conductivity, E_a. In general, the smaller well width, the higher E_a, thus in-plane electrical conductivity decreases as compared to the unlayered material. These effects are presented in a number of research papers, typical behaviour for diamond-like carbon films (DLC) superlattices, is shown in Figs. 20.3. and 20.4. However, the opposite situation may occur as well. This is because in amorphous semiconductors, unlike crystalline ones, there is a finite density of states within a mobility gap. When two such materials are put into contact, charge carriers may be transferred through an interface, building up accumulation and depletion layers, and as a result, the conductivity in one of the materials increases to such a degree that obscures quantum effects.

A completely different picture appears when conductivity in the direction perpendicular to the layers is considered. If there is no electric field, the current does not flow of course. When the small field is applied, a tunneling current can flow through the barriers. As the field is increased, an alignment between the energy subbands in adjacent wells is getting worse so observed maximum is a compromise between a driving force (electric field) and subbands mismatch.

When the field is increased further, the subbands misalignment is so high, that charge transport becomes very inefficient and the current—field characteristics exhibits minimum. If the field becomes higher, we can have the situation, that the lower subband in one quantum well is aligned with the upper one in the adjacent well. Transport is efficient again. The most important feature of this characteristics is a region of negative differential resistance where an increase of the electric field leads to a decrease of the electric current.

More detailed theoretical considerations of this matter can be found elsewhere. This kind of behaviour has been observed in both crystalline and amorphous multilayers; however, in the crystalline case it is more distinct and resembling model behaviour. An example of characteristics for DLC superlattice is shown, where current-voltage profile together with its first and second differentials is presented. Bumps observed on the current-voltage curve are typical of resonant tunneling. The start of the negative differential resistance region at 3.1 V corresponds to the loss of subbands alignment. It should be noted that the conductivity in the direction perpendicular to the superlattice layers is by its nature very similar to what Esaki found in *p-n* germanium junctions and what is now applied in construction of tunnel diodes. Typical *blue shift* effect is shown in Figs. 20.7 and 20.8.

The solid lines represents fit to a theoretical model of a one-dimensional quantum well with no coupling between energy levels of adjacent wells. We have discussed electrical and optical phenomena which characterise superlattice structures which found their explanation on the basis of quantum size effect. However, alternative interpretations of experimental data do exist as well. Arsenault et al. claim that bumps observed in the current-voltage characteristics for perpendicular conductivity in amorphous multilayers, usually attributed to resonant tunneling can be explained in a different way. They suggest that at least one of these bumps is simply a transition between two transport mechanisms: space charge limited current (low electric field) and multiple hopping (high field).

The most convincing justification of their hypothesis is that they observed the same bumps for both quantum well structures and single barriers. Quantum mechanical explanation of the *blue shift* has been criticised in a number of research papers as well. In our discussion it has been assumed that described multilayers were periodic system, characterized, from the geometrical point of view, only by one well and one barrier width. However it does not have to be the only approach. Another and very original view is presented in some researches. Superlattices of these researches consist of the layers of a-Ge and a-Si deposited using dual-target magnetron sputtering. The most interesting thing is their topology generated by Cantor sets. The modulation algorithm according to the so called "Cantor middle—third" set together with corresponding multilayer structure is presented.

SYNTHESIS ISSUES MEANS TO ACHIEVE MONODISPERSITY

One of the most challenging problems in synthesis is the controlled generation of monodispersed nanoparticles with size variance so small that size selection by centrifugal precipitation or mobility classification is not necessary. Among all the synthesis techniques discussed above, gas-phase synthesis is one of the best techniques with respect to size monodispersity, typically achieved by using a combination of rigorous control of nucleationcondensation growth and avoidance of coagulation by diffusion and turbulence as well as by the effective collection of nanoparticles and their handling afterwards. The stability of the collected nanoparticle powders against agglomeration, sintering, and compositional changes can be ensured by collecting the nanoparticles in liquid suspension.

For semiconducting particles, stabilization of the liquid suspension has been demonstrated by the addition of polar solvent (Murray et al. 1993); surfactant molecules have been used to stabilize the liquid suspension of metallic nanoparticles. Alternatively, inert silica encapsulation of nanoparticles by gas-phase reaction and by oxidation in colloidal solution has been shown to be effective for metallic nanoparticles (Andres et al. 1998). New approaches need to be developed for the generation of monodisperse nanoparticles that do not require the use of a size classification procedure. An example of this is a process developed in Japan where very monodispersed gold colloidal nanoparticles with diameters of about 1 nm have been prepared by reduction of metallic salt with UV irradiation in the presence of dendrimers (Esumi et al. 1998). Poly(amidoamine) dendrimers with surface amino groups of higher generations have spherical 3-D structures, which may have an effective protective action for the formation of gold nanoparticles. Although the specific role of dendrimers for the formation of monodispersed nanoparticles has yet to be defined, good monodispersity is thought to come from the complex reaction accompanying the decomposition of dendrimers, which eventually leads to the conversion of solution ions to gold nanoparticles. Scaleup production is of great interest for nanoparticle synthesis. High energy ball milling, already a commercial high-volume process, as mentioned above, has been instrumental in generating nanoparticles for the preparation of magnetic, structural, and catalytic materials. However, the process produces polydispersed amorphous powder, which requires subsequent partial recrystallization before the powder is consolidated into nanostructured materials. Although gas-phase synthesis is generally a low production rate process (typically in the 100 milligrams per hour range) in research laboratories, higher rates of production (about 20 grams per hour) are being demonstrated at Ångström Laboratory at Uppsala University in Sweden.

Even higher production rates (about 1 kg per hour) are now being achieved commercially. For sol-gel processing, the development of continuous processing techniques based on present knowledge of batch processing has yet to be addressed for economical scaleup production of nanoparticles. Other related sol-gel issues concern the cost of precursors and the recycling of solvent. Overall, sol-gel processing is attractive for commercial scale-up production. A recent paradigm shift envisioned for optoelectronics and computational devices involves the assembly of molecular or quantum wires (Chidsey and Murray 1986). Large polymeric molecules have been used as nano building blocks for nanoporous molecular sieves, biocompatible materials, optical switching, data processing, and other nonlinear optical components. Chain aggregates of nanoparticles can be considered as polymer-like units with their primary particles composed of a few hundred to a few thousand molecules. Thus, these chain aggregates can be considered "heavy" quantum wires. In fact, nanoparticle chain aggregates have been studied extensively as magnetic materials (Zhang and Manthiram 1997), as reinforced elastomers (Pu et al. 1997), and as additives in concrete (Sabir 1997). These aggregates have been shown to have chemical and mechanical properties different from those of individual primary particles (Friedlander et al. 1998). Depending on the particle size and its compositional material, the bonding force responsible for holding the aggregates together varies from weak van der Waals force for micrometer particles to strong chemical bonds for nanometer particles to very strong magnetic dipolar bonds for nanosized magnetic particles. The mechanical, optical, and electronic transport properties can be varied by controlling the diameter and the monodispersity of the primary particles, the crystalline structure and morphology, aggregate length, interfacial properties, and material purity. These chain aggregates can be formed by allowing agglomeration of nanoparticles generated by any of the synthesis techniques discussed above, with the exception of high energy ball milling, which generates particles with low surface area and high anisotropic morphologies, both of which are detrimental for the formation of chain aggregates. Depending on the magnetic and electric charging properties of the nanoparticles, an external applied magnetic or electric field can be used to control the fractal dimension of aggregates. For optical applications of chain aggregates, lower fractal dimensions (i.e., relatively straight chain aggregates with few branches) are desirable. Recent advances in the fabrication of nanometer fibers or tubes offer another form of building blocks for nanostructured materials. An effective way to generate nanometer fibers (or tubes) is based on the use of membrane-template techniques (Martin 1994). Membranes, with nanochannels generated by fission-fragment tracks or by electrochemical etching of aluminum metal, are used as templates for either chemical or electrochemical deposition of conductive polymers (Pathasarathy and Martin 1994), metal (van de Zande et al. 1997), semiconductor (Klein et al. 1993), and other materials for the generation of nanofibers or tubes.

Since the nanochannels on membranes are very uniform in size, the diameter and the aspect ratio of the nanofibers (or tubes) synthesized by the membranetemplate technique can be precisely controlled. This greatly facilitates the interpretation of optical data and the processing of these fibers (or tubes) into 2-D nanostructured materials (de Heer et al. 1995). Single-crystal semiconductor nanofibers can also be grown catalytically by metalorganic vapor phase epitaxy and laser ablation vapor-liquid-solid techniques (Hiruma et al. 1995; Morales and Lieber 1998). The synthesis of these onedimensional structures with diameters in the range of 3 to 15 nm holds considerable technological promise for optoelectronic device applications, such as the p-n junctions for light emission at Hitachi Central Research Laboratory in Japan. The advent of carbon-based nanotubes has created yet another way to fabricate nanometer fibers and tubes. These nanotubes have been used as templates for the fabrication of carbide and oxide nanotubes (Dai et al. 1995; Kasuga et al. 1998). Synthesis of nanotubes based on BN, BC3 and BC2N have also been reported (Chopra et al. 1995; Miyamoto et al. 1994). These nanotubes potentially possess large third-order optical non-linearity and other unusual properties (Xie and Jiang 1998). Metallic nanofibers synthesized by carbon-nanotube-template techniques are useful in the design of infrared absorption materials. The carbon nanotubes can now be catalytically produced in large quantities and have been used for reinforcement of nanostructural composite materials and concrete (Peigney et al. 1997). The elegant complexity of biological materials represents the achievement of structural order over many length scales, with the full structure developed from the "nested levels of structural hierarchy" (Aksay et al. 1992), in which self-assembled organic materials can form templates or scaffolding for inorganic components. These notions of a multilevel material structure with strong interactions among levels and an interplay of perfection and imperfection forming the final material was discussed earlier by Cyril Stanley Smith (1981). Along with characteristic length scales, there are characteristic relaxation times of the material, bringing in the consideration of the temporal stability of the structured materials (Zener 1948). A more detailed discussion of the role of biological materials as both paradigm and tool for the fabrication of nanostructured materials is given. It is interesting how many of the synthesis and assembly approaches seek out and adapt two key features of biogenic fabrication: that of "self-assembly," and the use of natural nanoscaled templates.

NANO BIOTECHNOLOGY: SELECT

Nanobiotechnologies have been applied to improve drug delivery and to overcome some of the problems of drug delivery in cancer. These can be classified into many categories that include use of various nanoparticles, nanoencapsulation, targeted delivery to tumors of various organs, and combination with other methods of treatment of cancer such as radiotherapy. Nanoparticles are also used for gene therapy for cancer. Some of the technologies enable combination of diagnostics with therapeutics which will be important for the personalized management of cancer. Some of the limitations of these technologies and prospects for future development are discussed.

Cancer drug delivery is no longer simply wrapping the drug in new formulations for different routes of delivery. Knowledge and experience from other technologies such as nanotechnology, advanced polymer chemistry, and electronic engineering, are being brought together in developing novel methods of drug delivery. Advances in our knowledge of molecular biology of cancer and pathways involved in malignant transformation of cells are revolutionizing the approach to cancer treatment with a focus is on targeted cancer therapy. The newer approaches to cancer treatment not only supplement the conventional chemotherapy and radiotherapy but also aim to prevent damage to the normal tissues and overcome drug resistance. Innovative methods of cancer treatment require new concepts of drug delivery in cancer.

Nanosized polyphosphazene-platinum (II) conjugates with a wide range of molecular weight from 24,000 to 115,000 were synthesized to study their tumor selectivity by enhanced permeability and retention (EPR) effect and their antitumor activity. It has been found from biodistribution study that the present polyphosphazene-Pt(II) conjugates exhibit high tumor selectivity by EPR effect with the tumor to tissue ratio (TTR) from 3.6 to 13 depending on the molecular size. These polymer conjugates have shown excellent in vivo antitumor activity against both murine and human cancer cell lines. In particular, xenograft trials of the conjugates have shown outstanding tumor inhibition effect on the stomach cancer cell line, YCC-3, which is one of the least responsive to the anticancer agents currently in clinical use, although the reason is not clearly explainable yet. The high in vivo activity seems to be attributed to the controlled-release of the antitumor active platinum (II) moiety, [GlyGluPt(dach] (dach=trans-(+/-)-1,2-diaminocyclohexane) from the phosphazene backbone by degradation in aqueous solution.

The aim of this work was to develop a formulation for Z-DEVD-FMK, a peptide which is a caspase inhibitor and has been used in experimental animal studies for a decade. Peptide loaded chitosan nanoparticles were obtained by ionotropic gelation process and Z-DEVD-FMK was quantified by an HPLC method. The influence of the initial peptide concentration on the nanoparticle characteristics and release behavior was evaluated. The CS nanoparticles have a particle diameter (Z-average) ranging from approximately 313-412 nm and a positive zeta potential (20-28 mV). The formulation with the initial peptide concentration of 400 ng/ml provided the highest loading capacity (0.46%) and the highest extent of release (65% at 24 h) suggesting the possibility to achieve a therapeutic dose. According to the data obtained, this chitosan-based nanotechnology opens new and interesting perspectives for anticaspase activity.

Cell internalisation and intracellular distribution of PEG-coated polyhexadecylcyanoacrylate (PEG-PHDCA) nanoparticles in rat brain endothelial cells (RBEC) have been investigated. A cell fractionation method has been developed based on the selective permeabilisation of RBEC plasma membrane by digitonin. By interacting with membrane cholesterol, digitonin creates pores allowing the release of soluble and diffusible species outside the cell. The selectivity of plasma membrane permeabilisation was controlled by using compartment markers such as lactate dehydrogenase (LDH) for cytoplasm and cathepsin B for lysosomes. An optimal digitonin concentration of 0.003% (w/v) has been identified to induce a pattern of membrane permeabilisation corresponding to the extraction of 72% LDH and less than 15% of Cathepsin B. Membrane permeabilisation at this digitonin concentration allows one to distinguish between the cell cytoplasm and its endo/ lysosomal fraction. This methodology was applied to investigate the intracellular distribution of the nanoparticles after their incubation with the RBEC. The results showed that PEG-PHDCA nanoparticles were able to be internalised to a higher extent than PHDCA nanoparticles (after 20 min incubation). Additionally, these nanoparticles displayed different patterns of intracellular capture, depending on their specific surface composition: PEG-PHDCA nanoparticles were 48% in the plasma membrane, 24% in the cytoplasm, 20% in vesicular compartments and 8% associated with the fraction of the nucleus, the cytoskeleton and caveolae suggesting that PEG-PHDCA nanoparticle uptake by RBEC is specific and presumably due to endocytosis. Confocal microscopy studies confirmed the cellular uptake of PEG-PHDCA nanoparticles.

The purpose of this study was to develop a novel lectin-conjugated isopropyl myristate (IPM)-incorporated PLGA nanoparticle system (NP) for the local delivery of paclitaxel to the lungs. Wheat germ agglutinin (WGA) was conjugated onto preformed IPM- and paclitaxel-loaded PLGA NPs by a two-step carbodiimide method following comparative uptake studies of Concanavalin A, Ricinus communis-120 and WGA on A549, H1299 and CCL-186 cells. WIT-NP with mean diameter of 331 nm and zeta potential of -4.3 mV were prepared with yield of 66% and paclitaxel encapsulation efficiency of 61%. Particle size was expanded by surface conjugation with WGA, while zeta potential was reduced by the addition of IPM and WGA. In vitro paclitaxel release profile was not affected by WGA but initial drug release was enhanced by adding IPM into the formulation. The

WIT-NP showed a burst-release of about 32% of the paclitaxel load within the first 5 h followed by a slow zero-order release of another 7% of the drug load in the next 115 h. Compared with the clinical paclitaxel formulation, paclitaxel-loaded nanoparticles without IPM or WGA, or paclitaxel-loaded nanoparticles with only IPM or WGA, the WIT-NP had superior in vitro cytotoxicity against A549 and H1299 cells. IC50 for WIT-NP after 5 and 24 h incubation with A549 cells were not significantly different (15.5 and 15 muM, respectively) whereas the clinical formulation was not cytotoxic after 5 h but had IC50 of 14 muM after 24 h incubation. WIT-NP exhibited stronger cell-killing effect because of more efficient cellular uptake via WGA-receptor-mediated endocytosis and IPM-facilitated release of paclitaxel from the NPs.

Therapeutic agents based on DNA or RNA oligonucleotides (e.g., antisense DNA oligonucleotide, small interfering RNA) require a regulation of their kinetics in cytoplasm to maintain an optimal concentration during the treatment period. In this respect, delivery of functional nanoparticles containing these drugs into cytoplasm has been thought to have a potential for the cytosolic controlled gene release. In this study, we establish a protocol for the encapsulation of nanoparticles into liposome, which is further fused with ultra violet-inactivated Sendai virus to compose fusogenic liposomes. When nanoparticles were encapsulated in conventional liposomes, endocytosis-mediated uptake of nanoparticles was observed. In contrast, numerous amounts of nanoparticles were delivered into the cytoplasm without any cytotoxicity when the particles were encapsulated in fusogenic liposomes. Additionally, fusogenic liposome showed a high ability to deliver nanoparticles containing DNA oligonucleotides into cytoplasm. These results indicate that this combinatorial nanotechnology using fusogenic liposome and nanoparticle is a valuable system for regulating the intracellular pharmacokinetics of gene-based drugs.

The intricate intracellular communication between stromal and epithelial cells, which involves cell-cell-, cell-insoluble extracellular matrix- (ECM), and cell-soluble factor-mediated signaling processes, is an attractive target for therapeutic intervention in hormone-refractory and bone-metastatic prostate cancer. In the present study we demonstrated that androgen-independent PC3 prostate cancer cells adhered to and migrated on vitronectin (VN), a major noncollagenous ECM in mature bone, through the expression of alphav-containing integrin receptors alphavbeta1 and alphavbeta5 on the cell surface, as determined by antibody function blocking assay and flow cytometry analysis. Small interfering RNAs (siRNAs) targeting human integrin alphav markedly reduced their respective mRNA and protein expression in cells, resulting in nearly complete reduction in VN-mediated cancer progression in vitro. In vivo quantitative bioluminescence analysis of human prostate cancer bone xenografts demonstrated for the first time that intratumoral administration of liposome-encapsulated human alphav-siRNAs significantly inhibits the growth of luciferase-tagged PC3 tumors in skeleton, which was associated with decreased integrin alphav expression and increased apoptosis in tumor cells. This integrin-based gene therapy is particularly suitable for the treatment of prostate cancer bone metastasis.

Paclitaxel is one of the best anticancer drugs, which has excellent therapeutic effects against a wide spectrum of cancers. The formulation of paclitaxel used in its currently clinical administration includes Cremophor EL, which has been found to cause serious side effects. Nanoparticle formulation of paclitaxel may provide an ideal solution for this problem and achieve a sustained chemotherapy. A novel copolymer, poly(lactide)-vitamin E TPGS (PLA-TPGS), was synthesized from lactide and d-alpha-tocopheryl polyethylene glycol 1000 succinate by bulk polymerization for nanoparticle formulation of anticancer drugs. (1)H NMR, FTIR and GPC were used to detect molecular structure of the copolymer. Paclitaxel-loaded PLA-TPGS nanoparticles were fabricated by a modified solvent extraction/evaporation technique with or without emulsifier involved, which were characterized by laser light scattering for size and size distribution; field emission scanning electron microscopy for surface morphology; zeta potential for surface charge; X-ray photoelectron spectroscopy for surface chemistry. The drug encapsulation efficiency and the in vitro drug release kinetics were measured by high-performance liquid chromatography. Formulation optimization was pursued. The particles were found of around 300nm in size and narrow size distribution. Of all, 89% drug encapsulation efficiency has been achieved for nanoparticles of 5% drug loading. The drug release from PLA-TPGS nanoparticles was found to be biphasic with an initial burst of 17% in the first day, followed by a sustained pattern with 51% release after 31 days.

To enhance the radiosensitivity of human colon cancer cells by docetaxel. METHODS: Immunoliposomal docetaxel was prepared by coupling monoclonal antibody against carcinoembryonic antigen to cyanuric chloride at the PEG terminus of liposome. LoVo adenocarcinoma cell line was treated with immunoliposomal docetaxel or/and irradiation. MTT colorimetric assay was used to estimate cytotoxicity of immunoliposomal docetaxel and radiotoxicity. Cell cycle redistribution and apoptosis were determined with flow cytometry. Survivin expression in LoVo cells was verified by immunohistochemistry. D801 morphologic analysis system was used to semi-quantify immunohistochemical staining of survivin. RESULTS: Cytotoxicity was induced by immunoliposomal docetaxel alone in a dose-dependent manner. Immunoli-posomal docetaxel yielded a cytotoxicity effect at a low dose of 2 nmol/L. With a single dose irradiation, the relative surviving fraction of LoVo cells showed a dose-dependent response, but there were no significant changes as radiation delivered from 4 to 8 Gy. Compared with liposomal docetaxel or single dose irradiation, strongly radiopotentiating effects of immunoliposomal docetaxel on LoVo cells were observed. A low dose of immunoliposomal docetaxel could yield sufficient radiosensitivity. Immunoliposomal docetaxel were achieved both specificity of the conjugated antibody and drug radiosensitization. Combined with radiation, immunoliposomal docetaxel significantly increased the percentage of G(2)/M cells and induced apoptosis, but significantly decreased the percentage of cells in G(2)/G(1) and S phase by comparison with liposomal docetaxel. Immunohistochemical analysis showed that the brown stained survivin was mainly in cytoplasm of LoVo cells. Semi-quantitative analysis of the survivin immunostaining showed that the expression of survivin in LoVo cells under irradiation with immunoliposomal docetaxel was significantly decreased. CONCLUSION: Immunoliposomal docetaxel is strongly effective for target radiosensitation in LoVo colon carcinoma cells, and may offer the potential to improve local radiotherapy.

We have prepared and screened a library of novel functionalized polymers for development of nanoparticle drug delivery systems. The polymer backbone consisting of two ester-linked, nontoxic, biological monomers, glycerol and adipic acid, was prepared using a hydrolytic enzyme. The specificity of the chosen enzyme yields a linear polymer with one free pendant hydroxyl group per repeat unit, which can be further functionalized. This protocol gives control over the backbone polymer molecular weight, together with the ability to incorporate various amounts of different fatty acyl substituents. These functionalized polymers are able to self-assemble into well-defined small particles of high homogeneity with a very low toxicity. They are able to incorporate a water soluble drug, dexamethasone phosphate, with a high efficiency and drug loading which varies with the polymer specification. The above characteristics strongly suggest that these polymers could be developed into useful nanoparticulate drug delivery systems.

Systemic gene delivery using cationic liposome-DNA complexes (LDCs) has been shown to elicit potent antitumor activity in mice with tumor metastases to the lungs. However, intravenous gene delivery for treatment of established cancer has not been evaluated previously in a spontaneous, large animal model. We therefore evaluated the safety, toxicity, and efficacy of intravenous gene delivery, using LDCs in dogs with established tumor metastases. Twenty dogs with chemotherapy-resistant osteosarcoma metastases to the lungs received a series of intravenous infusions of cationic liposomes and plasmid DNA encoding the canine interleukin-2 (IL-2) cDNA. Effects of intravenous gene delivery on immune activation, clinical and hematologic parameters, tumor responses, and survival times were assessed. We found that slow intravenous administration of IL-2 LDCs resulted in detectable IL-2 transgene expression in lung tissues of dogs. Repeated intravenous infusions of LDCs were well tolerated by dogs with lung tumor metastases and elicited systemic immune activation, as reflected by fever, leukogram changes, monocyte activation, and increased natural killer cell activity. Three of 20 dogs experienced partial or complete regression of lung metastases after infusion of IL-2 LDCs. Overall survival times were significantly increased in treated dogs compared with historical control animals with the same stage of disease. We conclude that repeated intravenous infusion of LDCs in cancerbearing dogs is safe and well tolerated at low doses and may be capable of eliciting antitumor activity in some animals with advanced tumor metastases.

An amphiphilic, hyperbranched polymer suitable for use in controlled drug delivery is reported. This polymer was obtained by modification of the hyperbranched aliphatic polyester Boltorntrade mark H20 (H20) with succinic anhydride and then glycidyl methacrylate, and formed nanoparticles in aqueous solution. The critical association concentration was 7.4 x 10(-3) g. L(-1), as determined by fluorescence spectroscopy using pyrene as a molecular probe. A static/dynamic laser light scattering (LLS) study revealed that the average particle size was 39.4 nm with a low particle size distribution (PDI = 0.04), and that each particle was composed of about 350 amphiphilic molecules. Daidzein, a hydrophobic traditional Chinese medicine, was encapsulated during particle formation and the release properties were determined. The optimal feeding concentration of daidzein to hyperbranched polyester was 4.9 x 10(-5) g. mL(-1) to 5.0 x 10(-3) g. mL(-1) with a loading efficiency of 76.1%. In the presence of the enzyme Lipase PS, the drug loaded nanoparticles degraded in a random one-by-one manner and released the drug over a few days. This system is therefore a novel controlled drug release system based on nanoparticles formed of hyperbranched polyester.Encapsulation of daidzein by hyperbranched polyester particles.

Pain, often considered the 5th vital sign, plays a significant role in the pathophysiology of cancer. Often addressed as an afterthought, untreated or under-treated cancer-related pain can have deleterious effects on a patient's physical and psychological well-being. Additionally, patients with breakthrough cancer pain tend to have more intense and more frequent background pain than patients without breakthrough pain. Currently, only oral transmucosal fentanyl citrate (OTFC) has reached the market for the treatment of cancer-related breakthrough pain. OTFC is an excellent first step in addressing the unmet need for symptomatic relief of breakthrough cancer pain; however, there is much room for improvement. Nanoscale science and engineering advancements have the long-term potential to bring revolutionary changes in society and across virtually all physical, biological and engineering disciplines, particularly medicine. Nanotechnology offers the potential to address multiple, major unmet problems in the diagnosis, treatment and symptom management of a large variety of diseases and conditions, including cancer. Nanotechnology can engender transformational progress in crucial aspects of the fight against cancer, spanning the continuum that ranges from prevention, to early detection, screening and monitoring, to innovative diagnostics and therapeutic modalities in the era of patient-centered, molecular medicine. Specifically, the authors will discuss their current research in the field of biomedical nanotechnology and its application to the management of breakthrough cancer pain.

Cationic lipid-peptide-DNA-complexes (LPDs) consisting of AH-Chol-liposomes and protamine-phosphodiester-oligonucleotide-particles (proticles) were introduced as carriers for antisense therapy. The LPDs were physically characterized, and a possible mechanism for adsorption of oligonucleotides (ODNs) was suggested. An increase in stability of ODNs against DNase I and serum nuclease digestion by these carriers was demonstrated. The hydrodynamic diameter increased after incubation with FCS which could be attributed to a protein coating of the particle surface. However, in cell culture medium lower particle sizes of the complexes occurred. In an antisense c-myc in vitro model, the effect of LPDs was tested using U937 cells. The C-MYC level was reduced after treatment of these antisense ODN carrier complexes. Furthermore, no changes in target mRNA concentration of the treated cells was found by reverse transcription and competitive multiplex-PCR.

We conducted a phase I trial to determine the safety and maximum tolerated dose (MTD) of non-pegylated liposome-encapsulated doxorubicin (Myocettrade mark; D-99) administered with weekly docetaxel in metastatic breast cancer (MBC) patients. PATIENTS AND METHODS: : Twenty-one patients with no prior chemotherapy for MBC received D-99 (60 or 50 mg/m(2)) intravenously (i.v.) on day 1 and escalating doses of docetaxel (25, 30 and 35 mg/m(2)) i.v. on days 1 and 8 in cohorts of three to six patients. Treatment cycles were repeated every 21 days for a maximum of six cycles. RESULTS: : The maximum tolerated dose (MTD) was 50 mg/m(2) of D-99 in combination with 25 mg/m(2) of weekly docetaxel. The most common grade 4 toxicity was neutropenia that occurred in 42 (41%) of treatment cycles, with 10 hospitalizations for febrile neutropenia. Serious protocol-defined cardiac events occurred in three (14%) patients, with two (10%; 95% confidence interval [CI] 1% to 30%) developing congestive heart failure (CHF) after a total cumulative anthracycline dose (adjuvant doxorubicin + D-99) of 540 mg/m(2). CONCLUSIONS: : D-99 in combination with weekly docetaxel, at the doses and schedule as administered in this trial, is not recommended for phase II testing. Additional trials, using different doses and schedules, are required to evaluate the potential side-effects and efficacy of D-99 and docetaxel.

Idarubicin has been successfully encapsulated in cholesterol-free liposomes, however, little is known about how the rate of drug release from circulating liposomes influences therapeutic activity. The studies described herein assess the attributes of a liposome formulation required to significantly increase the plasma levels of idarubicin and further establish whether increases in the circulation longevity of the drug mediate improved antitumor activity. Pharmacokinetic assessments of 6 different (3)[H]-labelled liposome formulations were compared to free idarubicin. The highest idarubicin plasma concentrations were observed with DSPC/DSPE-PEG(2000) liposomes formulated with 2 mol% DSPE-PEG(2000) and 150 mM (iso-osmotic) internal citrate concentration. It was shown that increased levels of PEG-lipid incorporation augmented IDA release and the optimal liposomal formulation needed to be prepared under iso-osmotic conditions. For efficacy studies in a murine leukemia model, groups of 12-14 mice were treated i.v. with saline or equivalent doses (1, 2, 3 mg/kg) of free or liposomal IDA. Liposomal treatment groups exhibited a higher % increase in life span (ILS) as compared to equivalent doses of free drug. Efficacy studies completed in two drug resistant models, P388/ADR and MDA435LCC6/MDR1, demonstrated that neither the free nor liposomal formulation of idarubicin was therapeutically active. Encapsulation of IDA in liposomes increased antitumor activity in an IDA sensitive model, however, the significant increase in plasma drug levels was not sufficient to overcome multidrug resistance.

The main objective was to develop an optimal paclitaxel microemulsion prepared by self-microemulsifying drug delivery system (SMEDDS) which is a mixture of paclitaxel, tetraglycol, Cremophor ELP, and Labrafil 1944 and a paclitaxel microemulsion containing poly(D,L-lactide-co-glycolide) (PLGA) in order to offer controlled release of paclitaxel. To achieve this goal, paclitaxel and PLGA were dissolved by solubilizer like tetraglycol. There was not observed any change in molecular weight of PLGA after being solubilized by tetraglycol. The droplet size for all of the formulation of microemulsion was found in the range of 45-270nm by dynamic light scattering (DLS). It was observed that the droplet size of microemulsion without PLGA was smaller than that of microemulsion containing PLGA by transmission electron microscopy (TEM). The droplet of microemulsion containing PLGA was almost of spherical shape with smooth surface and there was no aggregation or adhesion among droplet of microemulsion by atomic force microscopy (AFM). The release behaviour of paclitaxel from microemulsion containing PLGA having various molecular weights (8K, 33K, and 90K) exhibited a biphasic pattern characterized by a fast initial release during the first 48h, followed by a slower and continuous release for 144h, in contrast that the release of paclitaxel from microemulsion without PLGA was finished during 24h. This result was identical with the result of anti-tumor activity in vitro of paclitaxel from microemulsion containing PLGA against human breast cancer cell line MCF7 and this formulation enhanced anti-tumor activity in vivo compared with microemulsion without PLGA against SKOV-3 human ovarian cancer cells bearing nude mice model.

Chemotherapy remains the preferred choice of treatment for prostate cancer but modest drug response and significant toxicity by conventional methods of administration limit their efficacy. In our study, we determined the efficacy of paclitaxel (Tx)-loaded biodegradable nanoparticles (NPs) on tumor inhibition. We hypothesized that NPs following conjugation to transferrin (Tf) ligand (NPs-Tf) would enhance the therapeutic efficacy of the encapsulated drug. The antiproliferative activity of NPs was determined in human prostate cancer cell line (PC3) and their effect on tumor inhibition in a murine model of prostate cancer. NPs (approximately 220 nm in diameter, 5.4% w/w drug loading) under in vitro conditions exhibited sustained release of the encapsulated drug (60% release in 60 days). The IC50 (concentration of drug for 50% inhibition of cell growth) of the drug with Tf-conjugated NPs (Tx-NPs-Tf) was about 5-fold lower than that with unconjugated NPs (Tx-NPs) or drug in solution. Animals that received a single-dose intratumoral injection of Tx-NPs-Tf (Tx dose= 4 mg/kg) demonstrated complete tumor regression and greater survival rate than those that received either Tx-NPs or Tx-Cremophor EL formulation. In conclusion, sustained release NPs demonstrated greater antitumor activity following their conjugation to Tf ligand.

Infection can protect against subsequent disease by induction of both humoral and cellular immunity, but inert protein-based vaccines are not as effective. In this study, we present a new vaccine design, with Ag covalently conjugated to solid core nano-beads of narrowly defined size (0.04-0.05 microm) that localize to dendritic cells (DEC205(+) CD40(+), CD86(+)) in draining lymph nodes, inducing high levels of IFN-gamma production (CD8 T cells: precursor frequencies 1/5000 to 1/1000) and high Ab titers in mice. Conjugation of Ag to these nano-beads induced responses that were significantly higher (2- to 10-fold) than those elicited by other bead sizes, and higher than a range of currently used adjuvants (alum, QuilA, monophosphoryl lipid A). Responses were comparable to CFA/IFA immunization for Abs and ex vivo peptide-pulsed dendritic cell immunization for CD8 T cells. A single dose of Ag-conjugated beads protected mice from tumors in two different model challenges and caused rapid clearance of established tumors in mice. Thus, a range of Ags conjugated to nano-beads was effective as immunogens in both therapeutic and prophylactic scenarios.

The aim was to investigate the body distribution of nanoparticle-containing adriamycin (NADR) injected into the hepatic artery of hepatoma-bearing rats. Thirty Walker-256 hepatoma-bearing rats were divided into two groups at random, with 15 rats in each. NADR and free adriamycin (FADR) were injected into the hepatic artery of animals on the seventh day after tumor implantation. At 1, 5, and 15 hr, after administration, five animals in each group were sacrificed and the ADR concentrations in the plasma, liver, heart, spleen, lungs, kidneys, and tumor were determined. The results showed that NADR substantially increased the ADR concentrations in liver, spleen, and tumor of rats compared to FADR, whereas the concentrations in plasma, heart, and lungs were significantly decreased. In conclusion, the body distribution of ADR can be modified by its encapsulation into nanoparticles and administration via the hepatic artery.

To evaluate the effect of aclacinomycin A-loaded magnetic polybutylcyanoacrylate nanoparticles on gastric tumor growth in vivo and in vitro. Magnetic polybutylcyanoacrylate (PBCA) nanospheres encapsulated with aclacinomycin A (MPNS-ACM) were prepared by interfacial polymerization. Particle size, shape and drug content were examined. Female BABL/c nude mice were implanted with MKN-45 gastric carcinoma tissues subcutaneously to establish human gastric carcinoma model. The mice were randomly divided into 5 groups of 6 each: ACM group (8 mg/kg bm); group of high dosage of MPNS-ACM (8 mg/kg bm); group of low dosage of MPNS-ACM (1.6 mg/kg bm); group of magnetic PBCA nanosphere (MPNS) and control group (normal saline). Magnets (2.5 T) were implanted into the tumor masses in all of the mice one day before the therapy. Above-mentioned drugs were administered intravenously to the mice of every group on the first day and sixth day. When the mice were sacrificed, tumor weight was measured, and the assay of granulocyte-macrophage colony forming-unit (CFU-GM) was performed on semi-solid culture. White blood cell, alanine aminotransferase and creatine were examined. 3-(4-dimethylthiazol-2-yl)-2,5-diphenyltetrazolium bromide (MTT) was used to examine the viability of MKN-45 cells after incubation with different concentrations of ACM, MPNS and MPNS-ACM suspension respectively for 48 h. Content of ACM in MPNS-ACM was 12.0% and the average diameter of the particles was 210 nm. The inhibitory rates of ACM (8 mg/kg bm), high dosage of MPNS-ACM (8 mg/kg bm), low dosage of MPNS-ACM (1.6 mg/kg bm) and MPNS on human gastric carcinoma in nude mice were 22.63%, 52.55%, 30.66% and 10.22%, respectively. There was a significant decrease in the number of CFU-GM of bone marrow in ACM group compared with control group, whereas no obvious change was observed in that of the nanosphere groups. The values of 50% inhibition concentration (IC50) of ACM, MPNS and MPNS-ACM were 0.09, 97.78 and 1.07 microg/mL, respectively. The tumor inhibitory rate of MPNS-ACM was much higher than that of ACM under magnetic field and the inhibition on bone marrow was alleviated significantly compared with ACM group.

To investigate the antitumor activity of adriamycin (ADR) encapsulated in nanoparticles (NADR) and injected into the hepatic artery of hepatoma-bearing rats. NADR was prepared by the interfacial polymerization method. Walker-256 carcinosarcomas were surgically implanted into the left liver lobes of 60 male Wistar rats, which were divided into 4 groups at random (15 rats per group). On the 7th day after implantation, normal saline (NS), free ADR (FADR), NADR, or ADR mixed with unloaded nanoparticles (ADR+NP) was respectively injected via the hepatic artery (i.a.) of rats in different groups. The dose of ADR in each formulation was 2.0 mg/kg body weight and the concentration was 1.0 mg/mL. Survival time, tumor enlargement ratio, and tumor necrosis degree were compared between each group. Compared with the rats that received NS i.a., the rats that received FADR or ADR+NP acquired apparent inhibition on tumor growth, as well as prolonged their life span. Further significant anticancer efficacy was observed in rats that received i.a. administration of NADR. Statistics indicated that NADR brought on a more significant tumor inhibition and more extensive tumor necrosis, as compared to FADR or ADR+NP. The mean tumor enlargement ratio on the 7th day after NADR i.a. was 1.106. The mean tumor-bearing survival time was 39.50 days. Prolonged life span ratio was 109.22% as compared with rats that accepted NS. Therapeutic effect of ADR on liver malignancy can be significantly enhanced by its nanopaticle formulation and administration via hepatic artery.

The following study examines the feasibility of nanoshell-assisted photo-thermal therapy (NAPT). This technique takes advantage of the strong near infrared (NIR) absorption of nanoshells, a new class of gold nanoparticles with tunable optical absorptivities that can undergo passive extravasation from the abnormal tumor vasculature due to their nanoscale size. Tumors were grown in immune-competent mice by subcutaneous injection of murine colon carcinoma cells (CT26.WT). Polyethylene glycol (PEG) coated nanoshells (approximately 130 nm diameter) with peak optical absorption in the NIR were intravenously injected and allowed to circulate for 6 h. Tumors were then illuminated with a diode laser (808 nm, 4 W/cm2, 3 min). All such treated tumors abated and treated mice appeared healthy and tumor free >90 days later. Control animals and additional sham-treatment animals (laser treatment without nanoshell injection) were euthanized when tumors grew to a predetermined size, which occurred 6-19 days post-treatment. This simple, non-invasive procedure shows great promise as a technique for selective photo-thermal tumor ablation.

Biodegradable nanoparticles made of poly(D,L-lactic acid-co-glycolic acid) (PLGA) copolymer were characterized for enhanced delivery of antigens to murine bone marrow derived dendritic cells (DCs) in vitro. PLGA nanoparticles were efficiently phagocytosed by the DCs (CD11c+, MHC class II+, CD86+) in culture, resulting in their intracellular localization. The efficiency of the uptake was influenced by the incubation time and nanoparticle concentration. DCs pulsed with PLGA nanoparticles containing an immunomodulator, monophosphoryl lipid A(MPLA), showed upregulation of surface expression of MHC class II and CD86 molecules. Delivery of a cancer-associated antigen (MUC1 mucin peptide: BLP25) and MPLA in PLGA nanoparticles was shown to be superior to their delivery in the soluble form for activation of naive T cells of normal and MUC1-transgenic mice. These results strongly suggest that PLGA nanoparticles provide an efficient vaccine delivery system for targeting DCs and the development of DC based cellular vaccines.

Glioblastomas belong to the most aggressive human cancers with short survival times. Due to the blood-brain barrier, they are mostly inaccessible to traditional chemotherapy. We have recently shown that doxorubicin bound to polysorbate-coated nanoparticles crossed the intact blood-brain barrier, thus reaching therapeutic concentrations in the brain. Here, we investigated the therapeutic potential of this formulation of doxorubicin in vivo using an animal model created by implantation of 101/8 glioblastoma tumor in rat brains. Groups of 5-8 glioblastoma-bearing rats (total n = 151) were subjected to 3 cycles of 1.5-2.5 mg/kg body weight of doxorubicin in different formulations, including doxorubicin bound to polysorbate-coated nanoparticles. The animals were analyzed for survival (% median increase of survival time, Kaplan-Meier). Preliminary histology including immunocytochemistry (glial fibrillary acidic protein, ezrin, proliferation and apoptosis) was also performed. Rats treated with doxorubicin bound to polysorbate-coated nanoparticles had significantly higher survival times compared with all other groups. Over 20% of the animals in this group showed a long-term remission. Preliminary histology confirmed lower tumor sizes and lower values for proliferation and apoptosis in this group. All groups of animals treated with polysorbate-containing formulations also had a slight inflammatory reaction to the tumor. There was no indication of neurotoxicity. Additionally, binding to nanoparticles may reduce the systemic toxicity of doxorubicin. This study showed that therapy with doxorubicin bound to nanoparticles offers a therapeutic potential for the treatment of human glioblastoma.

Nanobodies are the smallest fragments of naturally occurring single-domain antibodies that have evolved to be fully functional in the absence of a light chain. Nanobodies are strictly monomeric, very stable, and highly soluble entities. We identified a nanobody with subnanomolar affinity for the human tumor-associated carcinoembryonic antigen. This nanobody was conjugated to Enterobacter cloacae beta-lactamase, and its site-selective anticancer prodrug activation capacity was evaluated. The conjugate was readily purified in high yields without aggregation or loss of functionality of the constituents. In vitro experiments showed that the nanobody-enzyme conjugate effectively activated the release of phenylenediamine mustard from the cephalosporin nitrogen mustard prodrug 7-(4-carboxybutanamido) cephalosporin mustard at the surface of carcinoembryonic antigen-expressing LS174T cancer cells. In vivo studies demonstrated that the conjugate had an excellent biodistribution profile and induced regressions and cures of established tumor xenografts. The easy generation and manufacturing yield of nanobody-based conjugates together with their potent antitumor activity make nanobodies promising vehicles for new generation cancer therapeutics.

We used polypropylenimine dendrimers for delivering a 31 nt triplex-forming oligonucleotide (ODN) in breast, prostate and ovarian cancer cell lines, using 32P-labeled ODN. Dendrimers enhanced the uptake of ODN by approximately 14-fold in MDA-MB-231 breast cancer cells, compared with control ODN uptake. Dendrimers exerted their effect in a concentration- and molecular weight-dependent manner, with generation 4 (G-4) dendrimer having maximum efficacy. A similar increase in ODN uptake was found with MCF-7 and SK-BR-3 (breast), LNCaP (prostate) and SK-OV-3 (ovarian) cancer cells. The dendrimers had no significant effect on cell viability at concentrations at which maximum ODN uptake occurred. [3H]Thymidine incorporation showed that complexing the ODN with G-4 significantly increased the growth-inhibitory effect of the ODN. Western blot analysis showed a significant 65% reduction of c-myc protein level in ODN-G-4 treated cells compared with that of ODN-treated/control cells. Gel electrophoretic analysis showed that ODN remained intact in cells even after 48 h of treatment. The hydrodynamic radii of nanoparticles formed from ODN in the presence of the dendrimers were in the range of 130-280 nm, as determined by dynamic laser light scattering. Taken together, our results indicate that polypropylenimine dendrimers might be useful vehicles for delivering therapeutic oligonucleotides in cancer cells.

Integrin alpha(v)beta(3) and vascular endothelial growth factor receptor 2 (Flk-1) have been shown to be involved in tumor-induced angiogenesis. Selective targeting of upregulated alpha(v)beta(3) and Flk-1 on the neovasculature of tumors is a novel antiangiogenesis strategy for treating a wide variety of solid tumors. In the studies described here, we investigated the potential therapeutic efficacy of two three-component treatment regimens using two murine tumor models. The treatment regimens used nanoparticle (NP) based targeting agents radiolabeled with (90)Y. The small molecule integrin antagonist (IA) 4-[2-(3,4,5,6-tetrahydropyrimidin-2-ylamino)ethoxy]-benzoyl-2-(5)-aminoeth ylsulfonylamino-beta-alanine, which binds to the integrin alpha(v)beta(3), and a monoclonal antibody against murine Flk-1 (anti-Flk-1 MAb) were used to target the NPs. Murine tumor models K1735-M2 (melanoma) and CT-26 (colon adenocarcinoma) were used to evaluate the treatment efficacy. In K1735-M2 and CT-26 tumors, a single treatment with IA-NP-(90)Y (14.2 microg/g IA, 5 or 6 microCi/g (90)Y) caused a significant tumor growth delay compared to untreated control tumors, as well as tumors treated with IA, IA-NP, and NP-(90)Y, respectively ($p < 0.025$, Wilcoxon test). In K1735-M2 tumors, a single treatment with anti-Flk-1 MAb-NP-(90)Y (0.36 microg/g anti-Flk-1 MAb, 5 microCi/g (90)Y) also caused a significant tumor growth delay ($p < 0.05$, Wilcoxon test) compared to untreated tumors, as well as tumors treated with anti-Flk-1 MAb, anti-Flk-1 MAb-NP, and conventional radioimmunotherapy with (90)Y-labeled anti-Flk mAb. Anti-CD31 staining showed a marked decrease in vessel density in tumors treated with anti-Flk-1 MAb-NP-(90)Y, which was associated with a high level of apoptotic death in these tumors, as shown by terminal deoxynucleotidyl transferase-mediated dUTP nick-end labeling staining. The present studies provide proof of principle that targeted radiotherapy works using different targeting agents on a nanoparticle, to target both the integrin alpha(v)beta(3) and the vascular endothelial growth factor receptor. These encouraging results demonstrate the potential therapeutic efficacy of the IA-NP-(90)Y and anti-Flk-1 MAb-NP-(90)Y complexes as novel therapeutic agents for the treatment of a variety of tumor types.

Gold glyconanoparticles (GNPs) have been prepared as new multivalent tools that mimic glycosphingolipids on the cell surface. GNPs are highly soluble under physiological conditions, stable against enzymatic degradation and nontoxic. Thereby GNPs open up a novel promising multivalent platform for biological applications. It has recently been demonstrated that specific tumor-associated carbohydrate antigens (glycosphingolipids and glycoproteins) are involved in the initial step of tumor spreading. A mouse melanoma model was selected to test glyconanoparticles as possible inhibitors of experimental lung metastasis. A carbohydrate-carbohydrate interaction is proposed as the first recognition step for this process. Glyconanoparticles presenting lactose (lacto-GNPs) have been used successfully to significantly reduce the progression of experimental metastasis. This result shows for the first time a clear biological effect of lacto-GNPs, demonstrating the potential application of this glyconanotechnology in biological processes.

Pituitary adenomas constitute the most frequent neuroendocrine pathology, comprising up to 15% of primary intracranial tumors. Current therapies for pituitary tumors include surgery and radiotherapy, as well as pharmacological approaches for some types. Although all of these approaches have shown a significant degree of success, they are not devoid of unwanted side effects, and in most cases do not offer a permanent cure. Gene therapy-the transfer of genetic material for therapeutic purposes-has undergone an explosive development in the last few years. Within this context, the development of gene therapy approaches for the treatment of pituitary tumors emerges as a promising area of research.

We begin by presenting a brief account of the genesis of prolactinomas, with particular emphasis on how estradiol induces prolactinomas in animals. In so doing, we discuss the role of each of the recently discovered growth inhibitory and growth stimulatory substances and their interactions in estrogen action. We also evaluate the cell-cell communication that may govern these growth factor interactions and subsequently promote the growth and survival of prolactinomas. Current research efforts to implement gene therapy in pituitary tumors include the treatment of experimental prolactinomas or somatomammotropic tumors with adenoviral vector-mediated transfer of the suicide gene for the herpes simplex type 1 (HSV1) thymidine kinase, which converts the prodrug ganciclovir into a toxic metabolite. In some cases, the suicide transgene has been placed under the control of pituitary cell-type specific promoters, like the human prolactin or human growth hormone promoters. Also, regulatable adenoviral vector systems are being assessed in gene therapy approaches for experimental pituitary tumors. In a different type of approach, an adenoviral vector, encoding the human retinoblastoma suppressor oncogene, has been successfully used to rescue the phenotype of spontaneous pituitary tumors of the pars intermedia in mice. We close the article by discussing the future of molecular therapies. We point out that although, gene therapy represents a key step in the development of molecular medicine, it has inherent limitations. As a consequence, it is our view that at some point, genetic therapies will have to move from exogenous gene transfer (i.e. gene therapy) to endogenous gene repair. This approach will call for radically new technologies, such as nanotechnology, whose present state of development is outlined.

A 23-amino-acid synthetic lytic peptide (Hecate) was

covalently linked to magnetite nanoparticles and the lytic peptide-bound nanoparticles were characterized by X-ray absorption near-edge structure spectroscopy, transmission electron microscopy, and electron diffraction. Investigation of magnetic properties with a superconducting quantum interference device (SQUID) magnetometer has shown a reduction in the saturation magnetization (Ms) of magnetite nanoparticles after binding with lytic peptide. An in vitro cell culture assay with breast cancer cell lines MDA-MB-435S revealed that the lytic peptide-bound magnetite nanoparticles were therapeutically active.

Paclitaxel is one of the best antineoplastic drugs found from nature in the past decades, which has been found effective against a wide spectrum of cancers including ovarian cancer, breast cancer, small and non small cell lung cancer, colon cancer, head and neck cancer, multiple myeloma, melanoma, and Kaposi's sarcoma. Like many other anticancer drugs, it has difficulties in clinical administration due to its poor solubility in water and most pharmaceutical reagents. In its current clinical application, an adjuvant called Cremophor EL has to be employed, which has been found to be responsible for many serious side effects. Nanoparticles of biodegradable polymers can provide an ideal solution to such an adjuvant problem and realize a controlled and targeted delivery of the drug with better efficacy and less side effects. With further development, such as particle size optimization and surface coating, nanoparticle formulation of paclitaxel can promote a new concept of chemotherapy to realize its full efficacy and to improve quality of life of the patients, which includes personalized chemotherapy, local chemotherapy, sustained chemotherapy, oral chemotherapy, chemotherapy across the blood-brain barrier, chemotherapy across the microcirculation barrier, etc. The present research proposes a novel formulation for fabrication of nanoparticles of poly(lactic-co-glycolic acid) (PLGA) by a modified solvent extraction/evaporation technique, in which natural emulsifiers, such as phospholipids, cholesterol and vitamin E TPGS are creatively applied to achieve high drug encapsulation efficiency, desired drug released kinetics, high cell uptake and high cytotoxicity. The nanoparticles composed of various recipes and manufactured under various conditions were characterized by laser light scattering (LLS) for size and size distribution, scanning electron microscopy (SEM) and atomic force microscopy (AFM) for morphological properties, X-ray photoelectron spectroscopy (XPS) and Fourier Transformation Infrared Spectroscopy (FTIR) for surface chemistry, zeta-potential for surface charge, and differential scanning calorimetry (DSC) for the thermogram properties. The drug encapsulation efficiency and the drug release kinetics under in vitro conditions were measured by high performance liquid chromatography (HPLC). It was found that these natural emulsifiers have great advantages for nanoparticle formulation of paclitaxel over the traditional macromolecular emulsifiers, such as polyvinyl alcohol (PVA). Nanoparticles of desired small size and narrow size distribution can be obtained. The drug encapsulation efficiency can be achieved as high as 100 %. The released kinetics can be made under control. The HT-29 cancer cell line experiment showed that after 24 hours of incubation, the cell mortality caused by the drug administered by such nanoparticle formulation could be more than 13 times higher than that caused by the free drug under similar conditions.

The aim was to determine the critical determinants in nanoparticle-mediated gene transfection. It was hypothesized that different formulation parameters could affect the nanoparticle characteristics and hence its gene transfection. Nanoparticles encapsulating plasmid DNA encoding for firefly luciferase were formulated using polylactide (PLA) and poly (D,L-lactide-co-glycolide) (PLGA) polymers of different compositions and molecular weights. A multiple-emulsion solvent-evaporation method with polyvinyl alcohol (PVA) as an emulsifier was used to formulate DNA-loaded nanoparticles. Gene expression of nanoparticles was determined in breast cancer (MCF-7) and prostate cancer (PC-3) cell lines. Nanoparticles formulated using PLGA polymer demonstrated greater gene transfection than those formulated using PLA polymer, and this was attributed to the higher DNA release from PLGA nanoparticles. Higher-molecular-weight PLGA resulted in the formation of nanoparticles with higher DNA loading, which demonstrated higher gene expression than those formulated with lower-molecular-weight PLGA. In addition, the nanoparticles with lower amount of surface-associated PVA demonstrated higher gene transfection in both the cell lines. Higher gene transfection with these nanoparticles was attributed to their higher intracellular uptake and cytoplasmic levels. Further study demonstrated that the molecular weight and the degree of hydrolyzation of PVA used as an emulsifier also affect the gene expression of nanoparticles. Results thus demonstrate that the DNA loading in nanoparticles and its release, and the surface-associated PVA influencing the intracellular uptake and endolysosomal escape of nanoparticles, are some of the critical determinants in nanoparticle-mediated gene transfection.

Metal nanoshells are a class of nanoparticles with tunable optical resonances. In this article, an application of this technology to thermal ablative therapy for cancer is described. By tuning the nanoshells to strongly absorb light in the near infrared, where optical transmission through tissue is optimal, a distribution of nanoshells at depth in tissue can be used to deliver a therapeutic dose of heat by using moderately low exposures of extracorporeally applied near-infrared (NIR) light. Human breast carcinoma cells incubated with nanoshells in vitro were found to have undergone photothermally induced morbidity on exposure to NIR light (820 nm, 35 W/cm2), as determined by using a fluorescent viability stain. Cells without nanoshells displayed no loss in viability after the same periods and conditions of NIR illumination. Likewise, in vivo studies under magnetic resonance guidance revealed that exposure to low doses of NIR light (820 nm, 4 W/cm2) in solid tumors treated with metal nanoshells reached average maximum temperatures capable of inducing irreversible tissue damage (DeltaT = 37.4 +/- 6.6 degrees C) within 4-6 min. Controls treated without nanoshells demonstrated significantly lower average temperatures on exposure to NIR light (DeltaT < 10 degrees C). These findings demonstrated good correlation with histological findings. Tissues heated above the thermal damage threshold displayed coagulation, cell shrinkage, and loss of nuclear staining, which are indicators of irreversible thermal damage. Control tissues appeared undamaged.

Heat shock proteins (HSPs) are recognized as significant participants in immune reactions. We previously reported that expression of HSP70 in response to hyperthermia, produced using our original magnetite cationic liposomes (MCLs), induces antitumor immunity. In the present study, we examine whether the antitumor immunity induced by hyperthermia is enhanced by hsp70 gene transfer. A human hsp70 gene mediated by cationic liposomes was injected into a B16 melanoma nodule in C57BL/6 mice in situ. At 24 hours after the injection of the hsp70 gene, MCLs were injected into melanoma nodules in C57BL/6 mice, which were subjected to an alternating magnetic field for 30 minutes. The temperature at the tumor reached 43 degrees C and was maintained by controlling the magnetic field intensity. The combined treatment strongly arrested tumor growth over a 30-day period, and complete regression of tumors was observed in 30% (3/10) of mice. Systemic antitumor immunity was induced in the cured mice. This study demonstrates that this novel therapeutic strategy combining the use of hsp70 gene therapy and hyperthermia using MCLs may be applicable to patients with advanced malignancies.

Hepatitis B virus envelope L particles form hollow nanoparticles displaying a peptide that is indispensable for liver-specific infection by hepatitis B virus in humans. Here we demonstrate the use of L particles for the efficient and specific transfer of a gene or drug into human hepatocytes both in culture and in a mouse xenograft model. In this model, intravenous injection of L particles carrying the gene for green fluorescent protein (GFP) or a fluorescent dye resulted in observable fluorescence only in human hepatocellular carcinomas but not in other human carcinomas or in mouse tissues. When the gene encoding human clotting factor IX was transferred into the xenograft model using L particles, factor IX was produced at levels relevant to the treatment of hemophilia B. The yeast-derived L particle is free of viral genomes, highly specific to human liver cells and able to accommodate drugs as well as genes. These advantages should facilitate targeted delivery of genes and drugs to the human liver.

Liposomes have been prepared with a polymerised core. Drug release and gastrointestinal (GI) degradation of liposomes with this polymerised core was slightly less important than those of classical liposomes. Vincristine, 5-fluorouracil (5-FU), and methotrexate (MTX) have been incorporated into the liposomes, and studies carried out using the differentiated cell lines Caco-2 and TC7, and with 150 histologically normal sections of human colon. Encapsulation of the drugs in liposomes had variable effects, depending on the test system and the drug used. For 5-FU and MTX calculated to be in a therapeutic range, liposomal formulation enhanced drug permeation, but not for the other drugs tested. In the excised human colon model, the treatment history of the patients can affect bioavailability: pre-operative radiation increased the drug tissue uptake. Transmucosal transport of ions was modified by prior chemotherapy. These results should be taken into account in the design of oral anticancer treatments both at the level of nutritional and pharmacological considerations.

A novel nanoparticle-based drug carrier for photodynamic therapy is reported which can provide stable aqueous dispersion of hydrophobic photosensitizers, yet preserve the key step of photogeneration of singlet oxygen, necessary for photodynamic action. A multidisciplinary approach is utilized which involves (i) nanochemistry in micellar cavity to produce these carriers, (ii) spectroscopy to confirm singlet oxygen production, and (iii) in vitro studies using tumor cells to investigate drug-carrier uptake and destruction of cancer cells by photodynamic action. Ultrafine organically modified silica-based nanoparticles (diameter approximately 30 nm), entrapping water-insoluble photosensitizing anticancer drug 2-devinyl-2-(1-hexyloxyethyl) pyropheophorbide, have been synthesized in the nonpolar core of micelles by hydrolysis of triethoxyvinylsilane. The resulting drug-doped nanoparticles are spherical, highly monodispersed, and stable in aqueous system. The entrapped drug is more fluorescent in aqueous medium than the free drug, permitting use of fluorescence bioimaging studies. Irradiation of the photosensitizing drug entrapped in nanoparticles with light of suitable wavelength results in efficient generation of singlet oxygen, which is made possible by the inherent porosity of the nanoparticles. In vitro studies have demonstrated the active uptake of drug-doped nanoparticles into the cytosol of tumor cells. Significant damage to such impregnated tumor cells was observed upon irradiation with light of wavelength 650 nm. Thus, the potential of using ceramic-based nanoparticles as drug carriers for photodynamic therapy has been demonstrated.

The cellular uptake, localization and efflux of meso-tetra-(4-hydroxyphenyl)porphyrin (p-THPP)-loaded nanoparticles have been studied in EMT-6 tumor cells. The effect of blood serum on photocytotoxicity has also been evaluated. Sub-130 nm nanoparticles based on poly(D,L-lactide-co-glycolide) (PLGA) (50:50 PLGA and 75:25 PLGA) and poly(D,L-lactide) (PLA) have been examined in comparison with free p-THPP. For all formulations tested, uptake of photosensitizer into cells was dependent on concentration, time and temperature.

All nanoparticulate formulations accumulated within the cells to a greater extent relative to free drug. Indeed, the fluorescence intensities measured on EMT-6 cells treated with p-THPP-loaded nanoparticulate formulations were at least two-fold higher than those obtained with free dye. Furthermore, the highest accumulation level was found with PLGA nanoparticles. Fluorescence microscopy revealed that endocytosis is a major intracellular sequestration mechanism of these p-THPP formulations and that these were localized into early and late endosomes.

The efflux study performed on both nonirradiated and irradiated cells indicated that free and p-THPP-loaded nanoparticles gradually escaped from EMT-6 cells as a function of time. This was more pronounced when cells were treated with nanoparticles and irradiated, reflecting important photodamage. It was also found that regardless of the nanoparticulate formulations tested, p-THPP photocytotoxicity was influenced by the concentration of the serum.

We describe a new method for selective cell targeting based on the use of light-absorbing microparticles and nanoparticles that are heated by short laser pulses to create highly localized cell damage. The method is closely related to chromophore-assisted laser inactivation and photodynamic therapy, but is driven solely by light absorption, without the need for photochemical intermediates (particularly singlet oxygen). The mechanism of light-particle interaction was investigated by nanosecond time-resolved microscopy and by thermal modeling. The extent of light-induced damage was investigated by cell lethality, by cell membrane permeability, and by protein inactivation. Strong particle size dependence was found for these interactions. A technique based on light to target endogenous particles is already being exploited to treat pigmented cells in dermatology and ophthalmology. With exogenous particles, phamacokinetics and biodistribution studies are needed before the method can be evaluated against photodynamic therapy for cancer treatment. However, particles are unique, unlike photosensitizers, in that they can remain stable and inert in cells for extended periods. Thus they may be particularly useful for prelabeling cells in engineered tissue before implantation. Subsequent irradiation with laser pulses will allow control of the implanted cells (inactivation or modulation) in a noninvasive manner.

Magnetic Drug Targeting means the specific delivery of chemotherapeutic agents to their desired targets, e.g. tumors, by using magnetic nanoparticles (ferrofluids) bound to these agents and an external magnetic field which is focused on the tumor. This type of target directed drug injection attempts to concentrate a pharmacologic agent by enhancing its efficacy while simultaneously minimizing deleterious side effects. In previous studies, we have been able to demonstrate the efficacy of this type of localized intraarterial chemotherapy in VX2 squamous cell carcinoma among rabbits [Alexiou, C., Arnold, W., Klein, R.J., Parak, F.G., Hulin, P., Bergemann, C., Erhardt, W., Wagenpfeil, S. and Lubbe, A.S. "Locoregional cancer treatment with Magnetic Drug Targeting", Cancer Res. 60 (2000) 6641-6648]. In the present investigation, we have studied the biodistribution of ferrofluids and chemotherapeutic agent by measuring the amount in the tumor, peritumoral area, various organs and body fluids (e.g. blood and urine), with and without Magnetic Drug Targeting. We compared results to that of administering a chemotherapeutic agent soley.

An external magnetic field was directed toward the tumor for 60 min. Biodistribution of ferrofluids in the tumor was investigated using histological cross sections and measured semi-quantitatively using 123I-labeled nanoparticles and quantitatively by the use of radioactive 59Fe-ferrofluids. Mitoxantrone was quantitatively measured using HPLC-analysis. The strength of the external magnetic field was 0.6 Tesla (permanent magnet) in the 123iodine study and 1.7 Tesla (electromagnet) in the 59Fe-study and HPLC-analysis.

The concentration of the ferrofluids (FFs) in the tumor region i.e. the tumor tissue and the surrounding area, which was under the influence of an external magnetic field, was found to be much higher than in the absence of one. In contrast to systemic chemotherapy, a much higher concentration of mitoxantrone in the tumor and the peritumoral area (region surrounding the tumor < or = 1 cm), by using only 50% and 20% of the normal dose was seen. Thus, the higher concentration of mitoxantrone could explain the therapeutic efficacy of Magnetic Drug Targeting in treatment of VX2 squamous cell carcinoma in rabbits in our previous studies with the advantage of no adverse clinical side effects.

Paclitaxel (Taxol) is one of the best antineoplastic drugs found from nature in the past decades. Like many other anticancer drugs, there are difficulties in its clinical administration due to its poor solubility. Therefore an adjuvant called Cremophor EL has to be employed, but this has been found to cause serious side-effects. However, nanoparticles of biodegradable polymers can provide an ideal solution to the adjuvant problem and realise a controlled and targeted delivery of the drug with better efficacy and fewer side-effects. The present research proposes a novel formulation for fabrication of nanoparticles of biodegradable polymers containing d-alpha-tocopheryl polyethylene glycol 1000 succinate (vitamin E TPGS or TPGS) to replace the current method of clinical administration and, with further modification, to provide an innovative solution for oral chemotherapy. In the modified solvent extraction/evaporation technique employed in this research, the emulsifier/stabiliser/additive and the matrix material can play a key role in determining the morphological, physicochemical and pharmaceutical properties of the produced nanoparticles. We found that vitamin E TPGS could be a novel surfactant as well as a matrix material when blended with other biodegradable polymers.

The nanoparticles composed of various formulations and manufactured under various conditions were characterised by laser light scattering (LLS) for size and size distribution, scanning electron microscopy (SEM) and atomic force microscopy (AFM) for morphological properties, X-ray photoelectron spectroscopy (XPS) for surface chemistry and differential scanning calorimetry (DSC) for thermogram properties. The drug encapsulation efficiency (EE) and the drug release kinetics under in vitro conditions were measured by high performance liquid chromatography (HPLC). It was concluded that vitamin E TPGS has great advantages for the manufacture of polymeric nanoparticles for controlled release of paclitaxel and other anti-cancer drugs. Nanoparticles of nanometer size with narrow distribution can be obtained. A drug encapsulation efficiency as high as 100% can be achieved and the release kinetics can be controlled.

The objective of this study was to target drug delivery to radiation-induced neoantigens, which include activated receptors within the tumor vasculature. These responses include posttranslational changes in pre-existing proteins, which can be discovered by phage-displayed peptide libraries administered to mice bearing irradiated tumors. Phage-displayed peptides recovered from irradiated tumors included the amino acid sequence RGDGSSV. This peptide binds to integrins within the tumor microvasculature. Immunohistochemical staining of irradiated tumors showed accumulation of fibrinogen receptor alpha(2b)beta(3) integrin. We studied tumor targeting efficiency of ligands to radiation-induced alpha(2b)beta(3). Radiopharmaceuticals were localized to irradiated tumors by use of alpha(2b)beta(3) ligands conjugated to nanoparticles and liposomes. Fibrinogen-conjugated nanoparticles bind to the radiation-activated receptor, obliterate tumor blood flow, and significantly increase regression and growth delay in irradiated tumors. Radiation-guided drug delivery to tumor blood vessels is a novel paradigm for targeted drug delivery.

Irinotecan-containing nanoparticles (NP) were prepared by coprecipitation with addition of water to acetone solution of poly(DL-lactic acid), poly(ethylene glycol)-block-poly(propylene glycol)-block-poly(ethylene glycol) and irinotecan, and subsequent evaporation of organic solvent.

NP were purified by gel filtration and used for experiments after condensation by evaporation. The obtained NP showed the drug content of 4.5% (w/w) and the mean particle diameter of 118 nm with the particle diameter distribution between 80-210 nm. When the antitumor effect was examined at a repeated dose of 20 mg irinotecan eq/kg for 3 d (3 x 20 mg/kg) using mice bearing Sarcoma 180 subcutaneously, only NP suppressed tumor growth significantly. After i.v. injection in rats, NP maintained irinotecan plasma concentration longer than CPT-11 aqueous solution. The present nanoparticle formation is suggested as a possibly useful dosage form of irinotecan against solid tumor.

The alpha(v)beta(3) integrin is a potential pharmacological target for anti-angiogenic therapy. A recent report describes the use of alpha(v)beta(3)-targeted nanoparticles to deliver a gene to tumour vasculature selectively. This resulted in substantial tumour regression in several experimental mouse tumour models. Hence, this approach has great potential for the treatment of human cancer.

In an attempt to increase the local concentration of tamoxifen in estrogen receptor positive breast cancer cells, we have prepared and characterized poly(epsilon-caprolactone) (PCL) nanoparticle formulation. METHODS: PCL (mol wt 14,800 daltons) nanoparticles were prepared by the solvent displacement method in acetone-water system in the presence of Pluronic F- 68. PCL nanoparticles, labeled with rhodamine123, were incubated with MCF-7 estrogen receptor positive breast cancer cells to determine uptake, intracellular distribution, and localization as a function of time. Intracellular drug concentrations over a specified period of time using different initial doses were examined using tritiated [3H]-tamoxifen. A significant fraction of the administered rhodamine123-loaded PCL nanoparticles was found in the perinuclear region of the MCF-7 cells, where estrogen receptors are also localized, after 1 hour of incubation. Measurements of the intracellular concentrations revealed that most of the administered nanoparticle dose was internalized within the first 30 minutes of incubation, and the uptake followed saturable transport kinetics. CONCLUSION: Results of this study show that PCL nanoparticles were rapidly internalized in MCF-7 cells and intracellular tamoxifen concentrations followed a saturable process. This approach may provide better therapeutic benefit by delivering the drug locally, near the tumor cells, for a longer period of time.

Dendritic cells (DCs) in the peripheral tissues act as sentinels of the immune system. They detect and capture pathogens entering the body and present their antigens to T cells to trigger responses directed towards elimination of the pathogen. The induction of peripheral tolerance against self and certain foreign antigens is also believed to be mediated through DCs. The outcome of any immune response is largely controlled by the microenvironment of antigen capture, processing and presentation by DCs. The "context" of antigen delivery to DCs will directly influence the microenvironment of antigen presentation and hence the regulation of immune responses. We report here preliminary investigations describing the formulation of a pharmaceutically acceptable, biodegradable, and strategic nanoparticulate delivery system, and its application for efficient antigen loading of DCs to achieve antigen specific T cell activation. "Pathogen-mimicking" nanoparticles capable of interacting with DCs were fabricated by incorporating monophosphoryl lipid A (MPLA; toll-like receptor (TLR) 4 ligand) or CpG ODN (seq #2006; TLR9 ligand) in biodegradable copolymer, poly(D,L,-lactic-co-glycolic acid) (PLGA). The uptake of PLGA nanoparticles by human umbilical cord blood derived DCs (DCs propagated from CD34 progenitors) was conclusively demonstrated by scanning electron microscopy (SEM), fluorescence activated cell sorting (FACS) and confocal laser scanning microscopy (CLSM). Cell phenotype at day 12 of cultures was determined as immature DC using specific cell surface markers, i.e. CD11c (approximately 90%), MHC-II (approximately 70%), CD86 (approximately 20%), CD83 (approximately 5%), CD80 (approximately 40%), CD40 (approximately 40%), and CCR7 (approximately 5%). Tetanus toxoid (TT), a model antigen, was encapsulated in nanoparticles along with an immunomodulator, i.e. either MPLA or CpG ODN. DCs pulsed with various antigen formulations were co-cultured with autologous naive T cells at various cell ratios (DC: T cells were 1:5-20). The DCs pulsed with TT and MPLA together in nanoparticles induced significantly higher T cell proliferation ($P<0.05$) as compared to when DCs pulsed with TT and MPLA in solution were employed. A similar trend was observed when CpG ODN was used instead of MPLA in the TT nanoparticles. This strategy of antigen delivery to DCs was then tested with a cancer vaccine candidate, a MUC1 lipopeptide. The T cell proliferation observed in the presence of nanoparticulate MUC1 and MPLA pulsed-DCs was much higher than DCs pulsed with soluble antigen ($P<0.0005$). These results indicate that PLGA nanoparticles mimicking certain features of pathogens are efficient delivery systems for targeting vaccine antigens to DCs and activation of potent T cell responses.

To increase the local concentration of tamoxifen in estrogen receptor (ER) positive breast cancer, we have developed and characterized nanoparticle formulation using poly(epsilon -caprolactone) (PCL). The nanoparticles were prepared by solvent displacement method using acetone-water system. Particle size analysis, scanning electron microscopy, zeta potential measurements, and differential scanning calorimetry (DSC) were used for nanoparticle characterization. Biodegradation studies were performed in the presence and absence of Pseudomonas lipase in phosphate-buffered saline (PBS, pH 7.4) at 37 degrees C. Tamoxifen loading over different concentrations was analyzed by high-performance liquid chromatography (HPLC) and the optimum loading concentration was determined. In vitro release studies were performed in 0.5% (w/v) sodium lauryl sulfate (SLS) containing PBS at 37 degrees C. Cellular uptake and distribution of fluorescent-labeled nanoparticles was examined in MCF-7 breast cancer cells. SEM micrographs and Coulter analysis showed nanoparticles with spherical shape and uniform size distribution (250-300 nm), respectively. Zeta potential analysis revealed a positive surface charge of +25 mV on the tamoxifen-loaded formulation. Being hydrophobic crystalline polyester, PCL did not degrade in PBS alone, but the degradation was enhanced by the presence of lipase. The maximum tamoxifen loading efficiency was 64%. Initial burst release of tamoxifen was observed, probably due to significant surface presence of the drug on the nanoparticles. A large fraction of the administered nanoparticle dose was taken up by MCF-7 cells through non-specific endocytosis. The nanoparticles were found in the perinuclear region after 1 h. Results of the study suggest that nanoparticle formulations of selective ER modulators, like tamoxifen, would provide increased therapeutic benefit by delivering the drug in the vicinity of the ER.

The aim of the present study was to evaluate the tumor accumulation of radiolabeled long-circulating poly(ethylene glycol) (PEG)-coated hexadecylcyanoacrylate nanospheres and non-PEG-coated hexadecylcyanoacrylate nanospheres (used as control), after intravenous injection in Fischer rats bearing intracerebrally well established 9L gliosarcoma. Both types of nanospheres showed an accumulation with a retention effect in the 9L tumor. However, long-circulating nanospheres concentrated 3.1 times higher in the gliosarcoma, compared with non-PEG-coated nanospheres. The tumor-to-brain ratio of pegylated nanospheres was found to be 11, which was in accordance with the ratios reported for other carriers tested for brain tumor targeting such as long-circulating liposomes or labels for magnetic resonance imaging. In addition, a 4- to 8-fold higher accumulation of the PEG-coated carriers was observed in normal brain regions, when compared with control nanospheres. Using a simplified pharmacokinetic model, two different mechanisms were proposed to explain this higher concentration of PEG-coated nanospheres in a tumoral brain. 1) in the 9L tumor, the preferential accumulation of pegylated nanospheres was attributable to their slower plasma clearance, relative to control nanospheres. Diffusion/convection was the proposed mechanism for extravasation of the nanospheres in the 9L interstitium, across the altered blood-brain barrier. 2) In addition, PEG-coated nanospheres displayed an affinity with the brain endothelial cells (normal brain region), which may not be considered as the result of a simple diffusion/ convection process. The exact underlying mechanism of such affinity deserves further investigation, since it was observed to be as important as specific interactions described for immunoliposomes with the blood-brain barrier.

The main objective of this study was to develop a polymeric drug delivery system for paclitaxel, intended to be intravenously administered, capable of improving the therapeutic index of the drug and devoid of the adverse effects of Cremophor EL. To achieve this goal paclitaxel (Ptx)-loaded poly(lactic-co-glycolic acid) (PLGA) nanoparticles (Ptx-PLGA-Nps) were prepared by the interfacial deposition method. The influence of different experimental parameters on the incorporation efficiency of paclitaxel in the nanoparticles was evaluated. Our results demonstrate that the incorporation efficiency of paclitaxel in nanoparticles was mostly affected by the method of preparation of the organic phase and also by the organic phase/aqueous phase ratio. Our data indicate that the methodology of preparation allowed the formation of spherical nanometric (<200 nm), homogeneous and negatively charged particles which are suitable for intravenous administration. The release behaviour of paclitaxel from the developed Nps exhibited a biphasic pattern characterised by an initial fast release during the first 24 h, followed by a slower and continuous release. The in vitro anti-tumoral activity of Ptx-PLGA-Nps developed in this work was assessed using a human small cell lung cancer cell line (NCI-H69 SCLC) and compared to the in vitro anti-tumoral activity of the commercial formulation Taxol. The influence of Cremophor EL on cell viability was also investigated. Exposure of NCI-H69 cells to 25 microg/ml Taxol resulted in a steep decrease in cell viability. Our results demonstrate that incorporation of Ptx in nanoparticles strongly enhances the cytotoxic effect of the drug as compared to Taxol, this effect being more relevant for prolonged incubation times.

The therapeutic profile of many anti-cancer drugs has been improved by their modified distribution through a colloidal carrier system. Hence, bovine serum albumin nanospheres containing 5-fluorouracil were prepared by pH-coacervation methods. To select the most suitable cryoprotector for the formulated nanosphere system, a study on the effect of cryoprotectors in the prevention of particle agglomeration was done. Using glucose and mannitol at various concentrations during freeze drying, glucose at a concentration of 5% was observed to be relatively more effective in the prevention of particle agglomeration than the other cryoprotectors. The carrier capacity was determined through the drug-to-albumin ratio. The particle size of all the drug-loaded batches was analyzed before and after freeze drying. The batch of nanospheres with uniform size distribution, and highest drug loading, was used for other subsequent studies. The effect of surfactant in drug loading was estimated through various concentrations of sodium lauryl sulfate, and it was observed that the surfactant has no influence on drug loading at the selected concentrations. The batch of nanospheres with highest drug loading was evaluated for its in-vitro release, and the drug release was found to be in a bi-phasic pattern. To evaluate the efficacy of 5-fluorouracil-loaded nanospheres against cancer cells, an in vitro cytotoxicity study was carried out using HEp-2 cell lines. The nanosphere-bound drug was observed to produce a better cytotoxic effect than the free drug. The anti-tumor efficacy of drug-loaded nanosphere was investigated in DLA tumor-induced mice models, and the percentage tumor inhibition was relatively higher in animals treated with nanosphere-bound drug than with free drug.

Gadolinium-incorporating lipid-nanoemulsions (Gd-nanoLE) for neutron-capture therapy were prepared. As a more convenient administration route than the intraperitoneal (i.p.) injection previously reported, the intravenous (i.v.) injections in tumor-bearing hamsters were carried out at an administration volume of 1 ml, which was the maximum tolerable injection volume of an i.v. injection and half that of an i.p. injection. When the standard-Gd-nanoLE of 1.5 mgGd/ml was administered, the absolute bioavailability in the i.p. injection was 57%, probably resulting from incomplete absorption from the peritoneal cavity into the blood stream. The biodistribution data revealed that the i.v. injection had three advantages over the i.p. injection, namely, a faster and higher accumulation of Gd-nano LE, and a more extended retention time in the tumor. Two i.v. injections of the standard-Gd-nanoLE with a 24h interval doubled the tumor accumulation of Gd, resulting in 49.7 microg Gd/g wet tumor 12h after administration. By using a twofold Gd-enriched formulation (High-Gd-nanoLE) of 3.0 mgGd/ml in the repeated administration schedule, the accumulation was doubled again, reaching 101 microg Gd/ g wet tumor. This level was comparable to the maximum level in the single i.p. injection previously reported. These results demonstrated that i.v. injection could be an alternative to i.p. injection as an administration route.

A mutant Raf-1 gene loaded onto nanoparticles, delivered to tumor vasculature with an integrin binding compound, provides effective antiangiogenic gene therapy in mice.

The objective of this study was to develop and characterize long-circulating, biodegradable, and biocompatible nanoparticulate formulation as an intracellular delivery vehicle. Poly(ethylene glycol) (PEG)-modified gelatin was synthesized by reacting Type-B gelatin with PEG-epoxide. The nanoparticles, prepared by pH and temperature controlled ethanol-water solvent displacement technique, were characterized for mean size, size distribution, and surface morphology. Electron spectroscopy for chemical analysis (ESCA) was used to confirm the surface presence of PEG chains. In vitro release of tetramethylrhodamine-labeled dextran (TMR-dextran, Mol. wt. 10,000 daltons) from the nanoparticle formulations was examined in PBS, with and without 0.2-mg/ml protease, at 37 degrees C. Relative cytotoxicity profile of control and PEGylated gelatin was evaluated in BT-20 a human breast cancer cell line. The nanoparticles were incubated with BT-20 cells to determine uptake and cellular distribution using confocal microscopy. Gelatin and PEGylated gelatin nanoparticles were found to be spherical in shape with a smooth surface in a size range of 200-500 nm and a unimodal size distribution. ESCA results showed an increase in the ether carbon (-C-O-) peak in the PEGylated gelatin nanoparticles due to the presence of PEG chains. The presence of PEG chains decreased the percent release of TMR-dextran in the presence of proteolytic enzyme due to steric repulsion. Cytotoxicity assays indicated that both gelatin and PEGylated gelatin were completely non-toxic to the cells. A large fraction of the administered control gelatin and PEGylated gelatin nanoparticles were found to be concentrated in the perinuclear region of the BT-20 cells after 12 hours indicating possible vesicular transport through initial uptake by endocytosis and endosomal processing. The results of this study show that PEGylation of gelatin may prove beneficial as long-circulating delivery system in vivo. Additionally, the nanoparticles could encapsulate hydrophilic macromolecules and are internalized by tumor cells.

Efforts to influence the biology of blood vessels by gene delivery have been hampered by a lack of targeting vectors specific for endothelial cells in diseased tissues. Here we show that a cationic nanoparticle (NP) coupled to an integrin alphavbeta3-targeting ligand can deliver genes selectively to angiogenic blood vessels in tumor-bearing mice. The therapeutic efficacy of this approach was tested by generating NPs conjugated to a mutant Raf gene, ATPmu-Raf, which blocks endothelial signaling and angiogenesis in response to multiple growth factors. Systemic injection of the NP into mice resulted in apoptosis of the tumor-associated endothelium, ultimately leading to tumor cell apoptosis and sustained regression of established primary and metastatic tumors.

Molecular therapy, including gene therapy, is a promising strategy for the treatment of human disease. However, delivery of molecular therapeutics efficiently and specifically to the target tissue remains a significant challenge. A human transferrin (Tf)-targeted cationic liposome-DNA complex, Tf-lipoplex, has shown high gene transfer efficiency and efficacy with human head and neck cancer in vitro and in vivo (Xu, L., Pirollo, K.F., Tang, W.H., Rait, A., and Chang, E.H. Hum. Gene Ther. 1999;10:2941-2952). Here we explore the structure, size, formation process, and structure-function relationships of Tf-lipoplex. We have observed Tf-lipoplex to have a highly compact structure, with a relatively uniform size of 50-90 nm. This nanostructure is novel in that it resembles a virus particle with a dense core enveloped by a membrane coated with Tf molecules spiking the surface. More importantly, compared with unliganded lipoplex, Tf-lipoplex shows enhanced stability, improved in vivo gene transfer efficiency, and long-term efficacy for systemic p53 gene therapy of human prostate cancer when used in combination with conventional radiotherapy. On the basis of our observations, we propose a multistep self-assembly process and Tf-facilitated DNA cocondensation model that may provide an explanation for the resultant small size and effectiveness of our nanostructural Tf-lipoplex system.

Cisplatin is one of the most widely used agents in the treatment of solid tumors, but its clinical utility is limited by toxicity. The development of less toxic, liposomal formulations of cisplatin has been hampered by the low water solubility and low lipophilicity of cisplatin, resulting in very low encapsulation efficiencies. We describe a novel method allowing the efficient encapsulation of cisplatin in a lipid formulation; it is based on repeated freezing and thawing of a concentrated solution of cisplatin in the presence of negatively charged phospholipids. The method is unique in that it generates nanocapsules, which are small aggregates of cisplatin covered by a single lipid bilayer. The nanocapsules have an unprecedented drug-to-lipid ratio and an in vitro cytotoxicity up to 1000-fold higher than the free drug. Analysis of the mechanism of nanocapsule formation suggests that the method may be generalized to other drugs showing low water solubility and lipophilicity.

Single cell manipulation, analytics, and label-free protein detection in microfluidic devices for systems nanobiology. Hellmich W, Pelargus C, Leffhalm K, Ros A, Anselmetti D., Electrophoresis. 2005 Sep 9.

Single cell analytics for proteomic analysis is considered a key method in the framework of systems nanobiology which allows a novel proteomics without being subjected to ensemble-averaging, cell-cycle, or cell-population effects. We are currently developing a single cell analytical method for protein fingerprinting combining a structured microfluidic device with latest optical laser technology for single cell manipulation (trapping and steering), free-solution electrophoretical protein separation, and (label-free) protein detection. In this paper we report on first results of this novel analytical device focusing on three main issues. First, single biological cells were trapped, injected, steered, and deposited by means of optical tweezers in a poly(dimethylsiloxane) microfluidic device and consecutively lysed with SDS at a predefined position. Second, separation and detection of fluorescent dyes, amino acids, and proteins were achieved with LIF detection in the visible (VIS) (488 nm) as well as in the deep UV (266 nm) spectral range for label-free, native protein detection. Minute concentrations of 100 fM injected fluorescein could be detected in the VIS and a first protein separation and label-free detection could be achieved in the UV spectral range. Third, first analytical experiments with single Sf9 insect cells (Spodoptera frugiperda) in a tailored microfluidic device exhibiting distinct electropherograms of a green fluorescent protein-construct proved the validity of the concept. Thus, the presented microfluidic concept allows novel and fascinating single cell experiments for systems nanobiology in the future.

The DNA delivery to mammalian cells is an essential tool for analyzing gene structure, regulation, and function. The approach holds great promise for the further development of gene therapy techniques and DNA vaccination strategies to treat and control diseases. Here, we report on the establishment of a cell-specific gene delivery and expression system by physical adsorption of a cell-recognition molecule on the nano-crystal surface of carbonate apatite. As a model, DNA/nano-particles were successfully coated with asialofetuin to facilitate uptake by hepatocyte-derived cell lines through the asialoglycoprotein receptor (ASGPr) and albumin to prevent non-specific interactions of the particles with cell-surface. The resulting composite particles with dual surface properties could accelerate DNA uptake and enhance expression to a notable extent. Nano-particles coated with transferrin in the same manner dramatically enhanced transgene expression in the corresponding receptor-bearing cells and thus our newly developed strategy represents a universal phenomenon for anchoring a bio-recognition macromolecule on the apatite crystal surface for targeted gene delivery, having immediate applications in basic research laboratories and great promise for gene therapy.

Bibliography

Anderlik, M. R. & Rothstein, M.A. Privacy and confidentiality of genetic information:What rules for the new science? *Annu. Rev. Genom.Hum.Genet.* 2, 401–433 (2001).

Aparicio, S. *et al.*Whole-genome shotgun assembly and analysis of the genome of *Fugu rubripes*. *Science* 297, 1301–1310 (2002).

Avery, O. T., MacLeod, C.M.& McCarty, M. Studies of the chemical nature of the substance inducing transformation of pneumococcal types. Induction of transformation by a desoxyribonucleic acid fraction isolated from *Pneumococcus* Type III. *J. Exp.Med.* 79, 137–158 (1944).

Banerjee, N. & Zhang, M.X. Functional genomics as applied to mapping transcription regulatory networks.*Curr.Opin.*

Bush, G.W., "Stem Cell Science and the Preservation of Life, " *The New York Times*, August 12, 2001, p. D13.

Cohen, S. N., Chang, A.C., Boyer, H.W. & Helling, R. B.

Collins, F. & Galas, D. A new five-year plan for the US Human Genome Project. *Science* 262, 43–46 (1993).

Collins, F. S. *et al.*New goals for the US Human Genome Project: 1998–2003. *Science* 282, 682–689 (1998).

Drews, J.& Ryser, S. The role of innovation in drug development.

Druker, B. J. Imatinib alone and in combination for chronic myeloid leukemia. *Semin.Hematol.* 40, 50–8 (2003).

Fuller, B. P. *et al.* Policy forum: Ethics — privacy in genetics research. *Science* 285, 1359–1361 (1999).

Gabriel, S. B. *et al.* The structure of haplotype blocks in the human genome. *Science* 296, 2225–2229 (2002).

Gardner, M. J. *et al.* Genome sequence of the human malaria parasite *Plasmodium falciparum*. *Nature* 419, 498–511 (2002).

Gavin, A. C. *et al.* Functional organization of the yeast proteome by systematic analysis of protein complexes. *Nature* 415, 141–147 (2002).

Golub, T.R.Genomic approaches to the pathogenesis of hematologic malignancy.*Curr.Opin.Hematol.* 8, 252–261 (2001).

Guttmacher and Mark S.Guyer are at the National Human Genome Research Institute, National Institutes of Health, Bethesda, Maryland 20892, USA.

Guttmacher, A. E.& Collins, F. S. Genomic medicine — A primer.*N.Engl. J.Med.* 347, 1512–1520 (2002).

Hannon, G. J.RNA interference. *Nature* 418, 244–251 (2002).

Hilbert, D.Mathematical problems. *Bull.Am.Math. Soc.* 8, 437–479 (1902).

Hirschhorn, J.N., Lohmueller, K., Byrne, E. & Hirschhorn, K.A comprehensive review of genetic association studies. *Genet.*

Holt, R. A. *et al.* The genome sequence of the malaria mosquito *Anopheles gambiae*. *Science* 298, 129–149 (2002).

Horii, T., Nagao, Y., Tokunaga, T., and Imai, H. 20003. Serum-free culture of murine primordial germ cells and embryonic germ cells. Theriogen. 59, 1257-1264.

Horii, T., Nagao, Y., Tokunaga, T., and Imai, H. 20003. Serum-free culture of murine primordial germ cells and embryonic germ cells. Theriogen. 59, 1257-1264.

Jackson, D. A., Symons, R. H. & Berg, P. Biochemical method for inserting new genetic information into DNA of Simian Virus 40: circular SV40 DNA molecules containing lambda phage genes and the galactose operon of *Escherichia coli*. *Proc.Natl Acad. Sci.*

Kaczorowski, D.J., Patterson, E.S., Jastromb, W.E., Shamblott, M.J. 2002. Review Article. Glucose-responsive insulin-producing cells from stem cells. Diabetes Metab.

Kaczorowski, D.J., Patterson, E.S., Jastromb, W.E., Shamblott, M.J. 2002. Review Article. Glucose-responsive insulin-producing cells from stem cells. Diabetes Metab.

Kalka, C. et al. Transplantation of ex vivo expanded endothelial progenitor cells for therapeutic neovascularization. Proc. Natl. Acad. Sci. USA 2000; 97:3422-3427.

Kalkman EAJ, Bilgin YM, van Haren P, van Suylen R-J, Saxena PR, Schoemaker RG.Determinants of coronary reserve in rats subjected to coronary artery ligation or aortic banding. Cardiovasc Res 1996.

Kawano H, Do YS, Kawano Y, Starnes V, Barr M, Law RE, Hsueh WA. Angiotensin II has multiple profibrotic effects in human cardiac fibroblasts.

Kehat I, Kenyagin-Karsenti D, Snir M, Segev H, Amit M, Gepstein A, Livne E, Binah O, Itskovitz-Eldor J, Gepstein L. Human embryonic stem cells can differentiate into myocytes with structural and functional properties of cardiomyocytes.

Kellerman S, Moore JA, Zierhut W, Zimmer HG, Campbell J, Gerdes AM. Nuclear DNA content and nucleation patterns in rat cardiac myocytes from different models of cardiac hypertrophy. *J Mol Cell Cardiol,* 497-505 (1992).

Kennedy M, Firpo M, Choi K, Wall C, Robertson S, Kabrun N, Keller G. A common precursor for primitive erythropoiesis and definitive haematopoiesis.

Kerr, D.A., Lladó, J., Shamblott, M.J., Maragakis, N.J., Irani, D.N., Crawford, T.O, Krishnan, C., Dike, S., Gearhart, J.D., and Rothstein, J.D. 2003. Human embryonic germ cell derivatives facilitate motor recovery of rats with diffuse motor neuron injury. J. Neurosci. 23, 5131-5140.

Kerr, D.A., Lladó, J., Shamblott, M.J., Maragakis, N.J., Irani, D.N., Crawford, T.O, Krishnan, C., Dike, S., Gearhart, J.D., and Rothstein, J.D. 2003. Human embryonic germ cell derivatives facilitate motor recovery of rats with diffuse motor neuron injury. J. Neurosci. 23, 5131-5140.

Kimura, T., Suzuki, A., Fujita, Y., Yomogida, K., Lomeli, H., Asada, N., Ikeuchi, M., Nagy, A., Mak, T.W., and Nakano, T. 2003. Development and Disease. Conditional loss of PTEN leads to testicular teratoma and enhances embryonic germ cell production. Development 130, 1691-1700.

Kimura, T., Suzuki, A., Fujita, Y., Yomogida, K., Lomeli, H., Asada, N., Ikeuchi, M., Nagy, A., Mak, T.W., and Nakano, T. 2003. Development and Disease. Conditional loss of PTEN leads to testicular teratoma and enhances embryonic germ cell production. Development 130, 1691-1700.

Klug MG, Soonpaa MH, Koh GY, Field LJ (1996) Genetically

selected cardiomyocytes from differentiating embryonic stem cells form stable intracardiac grafts. J Clin Invest 98:216-224.

Kocher A, Schuster M, Szabolcs M, Itescu S. Cardiomyocyte regeneration after neovascularization of ischemic myocardium by human bone marrowderived angioblasts. Nature Medicine 2002, in press.

Kocher AA, Schuster MD, Szabolcs MJ, Takuma S, Burkhoff D, Wang J, Homma S, Edwards NM, Itescu S. Neovascularization of ischemic myocardium by human bone-marrow-derived angioblasts prevents cardiomyocyte apoptosis, reduces remodeling and improves cardiac function. Nature Med.

L Field: Future therapy for cardiovascular disease. In Proceedings of the NHLBI Workshop Cell Transplantation: Future Therapy for Cardiovascular Disease? Columbia, MD; 1998.

Labastie M-C, Cortes F, Romeo P-H, Dulac C, Peault B. Molecular identity of hematopoietic precursor cells emerging in the human embryo. Blood 1998;92:3624-3635.

Levkau B, Koyama H, Raines EW, Clurman BE, Herren B, Orth K, Roberts JM, Ross R. Cleavage of p21cip1/waf1 and p27 kip1 mediates apoptosis in endothelial cells through activation of cdk2: role of a caspase cascade.

Li, Y., Jenkins, C.W., Nichols, M.A. and Xiong, Y. (1994) Cell cycle expression and p53 regulation of the cyclin-dependent kinase inhibitor p21. Oncogene, 9, 2261-2268 Steinman, R.A., Hoffman, B., Iro, A., Guillouf, C., Liebermann, D.A. and El- Houseini, M.E. (1994) Induction of p21 (WAF1/CIP1) during differentiation.

Liechty KW, MacKenzie TC, Shaaban AF, et al. Human

mesenchymal stem cells engraft and demonstrate site-specific differentiation after in utero transplantation in sheep. Nat Med 2000, 6:1282-6.

Lin Y, Weisdorf DJ, Solovey A, Hebbel RP. Origins of circulating endothelial cells and endothelial outgrowth from blood.

Lynch, H. T. & de la Chapelle, A. Genomic medicine: hereditary colorectal cancer.*N.Engl. J.Med.* 348, 919–932 (2003).

Lysaght, M.J., and Hazlehurst, A.L., "Private Sector Development of Stem Cell Technology and Therapeutic Cloning," *Tissue Engineering* 9(3): 555-561 (2003).

MacLellan WR and Schneider MD. Genetic dissection of cardiac growth control pathways Annu. Rev. Physiol. 2000. 62:289-320.

Mahon NG, O'Roke C, Codd MB, et al. Hospital mortality of acute myocardial infarction in the thrombolytic era. Heart 1999;81:478-82.

Maier S, Tertilt C, Chambron N, Gerauer K, Huser N, Heidecke C-D, Pfeffer K (2001) Inhibition of natural killer cells results in acceptance of cardiac allografts in CD28-/- mice. Nature Med 7:557-562.

Makino S, Fukuda K, Miyoshi S, et al. Cardiomyocytes can be generated from marrow stromal cells in vitro. J Clin Invest 1999, 103:697-705.

Maxam, A. M. & Gilbert, W. A new method for sequencing DNA.

McCarthy PM, Rose EA, Macmanus Q, et al. Clinical experience with the Novacor ventricular assist system. J Thorac Cardiovasc Surg. 1991; 102:578-587.

McEwan PE, Gray GA, Sherry L, Webb DJ, Kenyon CJ. Differential effects of angiotensin II on cardiac cell proliferation and intramyocardial perivascular fibrosis in vivo. Circulation 1998;98:2765-2773.

Mendel, G.Versuche über Pflanzen-Hybriden.*Verhandlungen des naturforschenden Vereines, Abhandlungen, Brünn* 4, 3–47 (1866).

Migeon, B.R., Chowdry, A.K., Dunston, J.A., and McIntosh, I. 2001. Identification of *TSIX*, encoding an RNA antisense to human *XIST*, reveals differences from its murine counterpart: Implications for X inactivation. Am. J. Human Genet. 69, 951-960.

Migeon, B.R., Chowdry, A.K., Dunston, J.A., and McIntosh, I. 2001. Identification of *TSIX*, encoding an RNA antisense to human *XIST*, reveals differences from its murine counterpart: Implications for X inactivation. Am. J. Human Genet. 69, 951-960.

Milki, A.A. et al., "Incidence of monozygotic twinning with blastocyst transfer compared to cleavage-stage transfer" *Fertility and Sterility* 79(3): 503-506 (2003).

Miller, P. S. Is there a pink slip in my genes? *J.Health Care Law Policy* 3, 225–265 (2000).

National Center for Health Statistics, "Births: Final Data for 2001." National Vital Statistics Reports 51(2) (2002), available at http://www.cdc.gov/nchs/data/ nvsr/ nvsr51/nvsr51_02.pdf.

National Institutes of Health Revitalization Act of 1993, Pub. L. No. 103-43, § 121(c), 107 Stat. 122 (1993), repealing 45 C.F.R. § 46.204(d).

National Institutes of Health, *Report of the Human Embryo Research Panel*, Bethesda, MD: NIH (1994).

National Research Council.*Mapping and Sequencing the*

Human Genome (National Academy Press, Washington DC, 1988).

Nelissen-Vrancken H, Debets J, Snoeckx L, Daemen M, Smits J. Time-related normalization of maximal coronary flow in isolated perfused hearts of rats with myocardial infarction. Circulation 1996;93:349-355.

Nidorf SM, Siu SC, Galambos G, Weyman AE, Picard MH. Benefit of late coronary reperfusion on ventricular morphology and function after myocardial infarction. J Am Coll Cardiol 1992;20:307-313.

Nirenberg, M.W. The genetic code: II. *Sci.Am.* 208, 80–94 (1963).

Norwitz, E. R., et al., "Implantation and the survival of early pregnancy," *New England Journal of Medicine* 345(19): 1400-1408 (2001)

Ogawa M, Kizumoto M, Nishikawa S, Fujimoto T, Kodama H, Nishikawa SI. Blood 1999;93:1168-1177.

Onyango, P., Jiang, S., Uejima, H., Shamblott, M.J., Gearhart, J.D., Cui, H., and Feinberg, A.P. 2002. Monoallelic expression and methylation of imprinted genes in human and mouse embryonic germ cell lineages. Proc. Natl. Acad. Sci. USA 99, 10599-10604.

Onyango, P., Jiang, S., Uejima, H., Shamblott, M.J., Gearhart, J.D., Cui, H., and Feinberg, A.P. 2002. Monoallelic expression and methylation of imprinted genes in human and mouse embryonic germ cell lineages. Proc. Natl. Acad. Sci. USA 99, 10599-10604.

Orlic D, Kajstura J, Chimenti S, et al. Bone marrow cells regenerate infarcted myocardium. Nature 2001, 410:701-705.

Oz MC, Argenziano M, Catanese KA, et al. Bridge experience with long-term implantable left ventricular assist devices. Are they an alternative to transplantation? Circulation. 1997; 95:1844-1852.

Park, T.S., Hong, Y.H., Kwon, S.C., Lim, J.M., and Han, J.Y. 2003. Birth of germline chimeras by transfer of chicken embryonic germ (EG) cells into recipient embryos.

Park, T.S., Hong, Y.H., Kwon, S.C., Lim, J.M., and Han, J.Y. 2003. Birth of germline chimeras by transfer of chicken embryonic germ (EG) cells into recipient embryos.

Pfeffer JM, Pfeffer MA, Fletcher PJ, Braunwald E. Progressive ventricular remodeling in rat with myocardial infarction. Am J Physiol 1991.

Pfeffer MA, Braunwald E, Moye LA, et al. Effect of captopril on mortality and morbidity in patients with left ventricular dysfunction after myocardial infarction. Results of the survival and ventricular enlargement trial. The SAVE investigators. N Engl J Med 1992;327:669-677.

Pitt B, Segal R, Martinez FA, et al. Randomised trial of losartan versus captopril in patients over 65 with heart failure (Evaluation of Losartan in the Elderly Study, ELITE). Lancet 1997;349:747-752.

Pittenger MF, Mackay AM, Beck SC, et al. Multilineage potential of adult human mesenchymal stem cells. Science 1999, 284:143-147.

Pouzet B, Ghostine S, Vilquin J-T, Garcin I, Scorsin M, Hagege AA, Duboc D, Schwartz K, Menasche P (2001) Is skeletal myoblast transplantation clinically relevant in the era of angiotensin-converting enzyme inhibitors? Circulation 104:I223-228.

Rafii S, Shapiro F, Rimarachin J, Nachman R, Ferris B, Weksler B, Moore AS, Asch AS. Isolation and characterization of human bone marrow microvascular endothelial cells: hematopoietic progenitor cell adhesion.

Ravichandran LV and Puvanakrishnan R. In vivo labeling studies on the biosynthesis and degradation of collagen in experimental myocardial myocardial infarction. Biochem Intl 1991;24:405-414.

Reich, D. E. & Lander, E. S. On the allelic spectrum of human disease. *Trends Genet.* 17, 502–510 (2001).

Rothenberg, K. *et al.* Genetic information and the workplace: Legislative approaches and policy challenges. *Science* 275, 1755–1757 (1997).

Sachidanandam, R. *et al.* A map of human genome sequence variation containing 1.42 million single nucleotide polymorphisms. *Nature* 409, 928–933 (2001).

Sanger, F.& Coulson, A. R. A rapid method for determining sequences in DNA by primed synthesis with DNA polymerase.

Sapienza, C. 2002. Commentary. Imprinted gene expression, transplantation medicine, and the "other" human embryonic stem cell. Proc. Natl. Acad. Sci. USA 99, 10243-10245.

Sapienza, C. 2002. Commentary. Imprinted gene expression, transplantation medicine, and the "other" human embryonic stem cell. Proc. Natl. Acad. Sci. USA 99, 10243-10245.

Selkoe, D. J.Alzheimer's disease: genes, proteins, and therapy. *Physiol. Rev.* 81, 741–66 (2001).

Shamblott, M.J., et al., "Derivation of pluripotent stem cells from cultured human primordial germ cells"

Proceedings of the National Academy of Sciences. 95(23): 13726-13731 (1998).

Shi Q, Rafii S, Wu MH-D, et al. Evidence for circulating bone marrow-derived endothelial cells. Blood 1998;92:362-367.

Sidow, A. Sequence first.Ask questions later. *Cell* 111, 13–16 (2002).

Smith, L. M. *et al.* Fluorescence detection in automated DNAsequence analysis. *Nature* 321, 674–679 (1986).

Smythe GM, Grounds MD (2000) Exposure to tissue culture conditions can adversely affect myoblast behaviour in vivo in whole muscle grafts: implications for myoblast transfer therapy. Cell Transplant 9: 379-393.

Smythe G.M., Hodgetts S.I., Grounds M.D. (2001) Problems and solutions in myoblast transfer therapy. J Cell Mol Med 5:33-47.

Stockwell, B. R. Chemical genetics: Ligand-based discovery of gene function. *Nature Rev. Genet.* 1, 116–125 (2000).

Takahashi, T. et al. Ischemia- and cytokine-induced mobilization of bone marrowderived endothelial progenitor cells for neovascularization. Nat. Med. 1999; 5:434-438.

The Chipping Forecast II. *Nature Genet.* 32, 461–552 (2002).

The Mouse Genome Sequencing Consortium. Initial sequencing and comparative analysis of the mouse genome. *Nature* 420, 520–562 (2002).

Tyson, J.J., Chen, K. & Novak, B.Network dynamics and cell physiology.*Nature Rev.Mol. Cell Biol.* 2, 908–916 (2001).

Van der Weyden, L., Adams, D. J. & Bradley, A. Tools for

targeted manipulation of the mouse genome. *Physiol.Genomics* 11, 133–164 (2002).

Wagner, K.R. Genetic diseases of muscle.*Neurol. Clin.* 20, 645–678 (2002).

Watson, J.D. & Crick, F.H. C. Molecular structure of nucleic acids: A structure for deoxyribose nucleic acid. *Nature* 171, 737 (1953).

Wilcox, A.J., et al., "Incidence of early loss of pregnancy," *New England Journal of Medicine* 319 (4): 189-194 (1988).

Xu, R.H., et al., "BMP4 initiates human embryonic stem cell differentiation to trophoblast" *Nature Biotechnology* 20(12): 1261-1264 (2002).

Zhang, M.Q.Computational prediction of eukaryotic proteincoding genes. *Nature Rev. Genet.* 3, 698–709 (2002).

Index